Health and Safety . . . in brief

Health and Safety
... in brief

John Ridley

Butterworth-Heinemann
225 Wildwood Avenue, Woburn, MA 01801-2041
Linacre House, Jordan Hill, Oxford OX2 8DP
A division of Reed Educational and Professional Publishing Ltd

A member of the Reed Elsevier plc group

OXFORD BOSTON JOHANNESBURG
MELBOURNE NEW DELHI SINGAPORE

First published 1998
Revised reprint 1998

© Reed Educational and Professional Publishing Ltd 1998

All rights reserved. No part of this publication may be reproduced in any material form (including photocopying or storing in any medium by electronic means and whether or not transiently or incidentally to some other use of this publication) without the written permission of the copyright holder except in accordance with the provisions of the Copyright, Designs and Patents Act 1988 or under the terms of a licence issued by the Copyright Licensing Agency Ltd, 90 Tottenham Court Road, London, England W1P 9HE. Applications for the copyright holder's written permission to reproduce any part of this publication should be addressed to the publishers

British Library Cataloguing in Publication Data
A catalogue record for this book is available from the British Library

Library of Congress Cataloguing in Publication Data
A catalogue record for this book is available from the Library of Congress

ISBN 0 7506 3765 X

Composition by Genesis Typesetting, Rochester, Kent
Printed and bound in Great Britain

Contents

Preface	ix
Part 1 Law	1
1 Legal processes	3
1.1 Historical background	3
1.2 Branches of law	4
1.3 Types of court	6
1.4 Court procedures	10
1.5 Law-making in UK	13
1.6 Law-making in the EU	15
2 Health and safety laws	18
2.1 A historical perspective on health and safety laws	19
2.2 The structure of health and safety laws	20
2.3 The Health and Safety at Work, etc. Act 1974	21
2.4 Health and safety law enforcement	22
2.5 The Management of Health and Safety at Work Regulations 1992	25
2.6 Other current health and safety legislation	29
Part 2 Management	31
3 Management responsibilities	33
3.1 The role of management	33
3.2 Responsibilities for health and safety	34
3.3 Safety policy	36
3.4 Risk assessments	38
3.5 Techniques of hazard identification	44
3.6 Organization for safety at work	46
3.7 Promoting health and safety in the workplace	48
4 Human resources	53
4.1 Health and safety training	53
4.2 Young persons at work	55
4.3 Joint consultation	59
4.4 Industrial relations in health and safety	61
4.5 Human factors in health and safety	63
4.6 Insurance in health and safety	65

5	Workplace safety	69
	5.1 Workplace Regulations	69
	5.2 Office safety	79
	5.3 Workplace safety signs and signals	82
6	Information and advice	87
	6.1 Safety advice	87
	6.2 Sources of information	88
	6.3 Report writing	91
	6.4 Data storage and retrieval	92
7	Accidents	94
	7.1 Principles of accident prevention	94
	7.2 Accident investigation	97
	7.3 Accident reporting	99

Part 3 Occupational health — 103

8	The body	105
	8.1 Functions of the body	105
	8.2 Routes of entry	109
	8.3 Target organs	110
9	Health at work	111
	9.1 Causes of health hazards	111
	9.2 Ionizing radiations	114
	9.3 Hazards from non-ionizing radiations	116
10	Health protection	117
	10.1 First aid	117
	10.2 Personal protective equipment	119
	10.3 Safe use of display screen equipment	122

Part 4 Safety technology — 125

11	Chemicals	127
	11.1 Safe use of chemicals	127
	11.2 Labelling of chemicals for supply and use	133
	11.3 Transport of chemicals by road and rail	137
	11.4 Classification of hazardous and dangerous substances for supply	141
	11.5 Approved lists	142
	11.6 Exposure limits	149
	11.7 Preventative and control measures	151
	11.8 Handling hazardous and dangerous substances	152
	11.9 Disposal of special waste	155
12	Noise and hearing protection	158
	12.1 Legislation concerning noise	158
	12.2 The ear	161
	12.3 Noise measurement	162
	12.4 Noise control techniques	164

13	Work equipment	167
	13.1 New machinery	167
	13.2 Safe use of work equipment	171
	13.3 Safety with moving machinery	176
	13.4 Safety during maintenance	179
	13.5 Pressure systems	183
	13.6 Lifting equipment	186
14	Construction	188
	14.1 CDM	188
	14.2 Construction health, safety and welfare	191
	14.3 Construction safety	195
	14.4 Employing contractors	198
	14.5 Access equipment	200
	14.6 Safety in demolition	203
	14.7 Safety with excavations	206
15	Manual handling	209
	15.1 The Manual Handling Operations Regulations 1992	209
	15.2 Safe manual handling	210
16	Mechanical handling	214
	16.1 Lifts	214
	16.2 Cranes	219
	16.3 Conveyors	221
	16.4 Powered trucks	222
17	Safe use of electricity	226
	17.1 The Electricity at Work Regulations 1989	226
	17.2 Safe use of electricity	229
	17.3 Safe use of portable electrical equipment	233
18	Fire	235
	18.1 Fire legislation	235
	18.2 Causes of fires and precautions	238
	18.3 Fire-fighting and extinguishers	240
	18.4 Safe use of flammable substances	243
19	Environment	245
	19.1 The Environmental Protection Act 1990	245
	19.2 Safe and healthy working environment	247
Appendix 1	List of statutes	249
Appendix 2	Abbreviations	252
Index		255

Preface

The text of this book started life as a set of notes I used for lectures on health and safety which became popular with the students as *aides-mémoire* for their revision prior to examinations. The students came from various positions in commerce and industry, ranging from union appointed safety representatives through middle managers to directors. Feedback from them indicated that they found the notes useful in their day-to-day duties. If they could, why couldn't others – and so the book was born.

In preparing this book I have endeavoured to take some of the mystique out of health and safety in the workplace and in simple plain English and in practical terms, explain the hows, whys and wherefores of keeping on the right side of health and safety laws.

It is not intended to be a substitute for the law. Readers must understand that when the crunch comes it is the responsibility of the employer/manager to ensure compliance with legislative requirements and, in the ultimate, for the courts to decide if compliance has been achieved. It is hoped that this book will help those responsible to meet their obligations and avoid running foul of the law.

There can be no doubt, the subject is vast and complex and to meet the objective of simplifying health and safety laws, their contents have been polarized into major components and the overall subject broken down into manageable parts. The text is purposefully kept brief – making use of 'bullet points' rather than verbiage – with references to sources of information should a particular situation demand more detailed study.

The text is restricted mainly to matters that are likely to be met in the day-to-day work of a busy manager with safety responsibilities, an emergent safety adviser, safety representative or those participating in safety awareness courses. While it may be of practical help in the design and manufacture of equipment, designers and manufacturers should refer to the various legislation and standards that relate to their particular product. This latter point is especially important within the context of the European Union.

With the written word it is not possible to include the inflexions of speech or the communication of body language so a convention has been adopted to differentiate between what is law and what is advice:

titles of and quotations from legislation are in	*italics*
extracts from laws are in	boxes with shading and indicated by a symbol in the left hand margin
advice and comments are in	brackets or normal type face.

I have also used abbreviations when referring to specific sections of an Act as 's.--' and to the clauses of a Regulation as 'r.--'

The advice and comments given draw on my experiences through the years of involvement in the drafting, enforcing and complying with various aspects of health and safety legislation. They are not intended to be comprehensive or exclusive but aim to draw attention to some of the essential features of the law and of current practices so that the individual, be he manager or safety adviser, can with reasonable confidence, and in the culture of his particular organization, make decisions on how to proceed or what action to take. Nor are they intended to denigrate legislative requirements, but to explain them in a simple manner so the task of complying is made easier for those working at the 'sharp end' of industry and commerce. Where a particular problem is met, reference should be made to the relevant legislation and published guidance on the subject.

The book is in four parts dealing with:

- the background to health and safety law
- employer's (and manager's) role and responsibilities
- health aspects
- technical aspects

Specific laws with health and safety connotations are dealt with where they arise in the context of the particular matter being dealt with, rather than including them in a general law section.

As far as possible the text has been written in the impersonal, but in places it has not been possible to do this so I have referred to *he*, *him* and *his*. This is not sexist but for convenience and I fully recognize and acknowledge the part that women play at all levels in our work lives. So wherever *he*, *him* or *his* occur please also read *she*, *her* or *hers*. I hope female readers will forgive me this convenience.

The contents cover in outline the syllabus for the National General Certificate level examination of the National Examination Board in Occupational Safety and Health and of the requirements for the Working Safely and Managing Safely Certificates sponsored by the Institution of Occuptional Safety and Health. It should also prove a useful *aide-mémoire* for those studying for higher level health and safety examinations.

To save repeating long titles, especially where they occur frequently in the text, I have used initials and have listed them in an appendix. There is also a comprehensive index to enable particular subject matter to be located easily and eliminate the time wasting task of wading through pages of text to find the pertinent information.

Health and safety laws are not static but are developing all the time to meet changing conditions and technologies. This situation is exacerbated by our committment to membership of the European Union which is increasingly dictating the subject matter of our laws.

The subject range of this book is vast and I have been grateful for the help, assistance and encouragement given to me in its preparation by Robin Rispin and Ray Chalklen. I would also like to thank Stocksigns for their help in providing copies of the hazard signs and John Stather of NRPB for his comments on Table 9.1.

Health and safety is not a serious subject in the sense that it is *dreary*, but it is serious in the sense that it is *important*, and as such is something that can be enjoyed and laughed at. So, to leaven the text, I have included a number of cartoons which I hope you will enjoy but which will also help you to remember particular points. I am grateful to Philip Wilson for those cartoons and showing us a lighter side to some of the stock safety phrases we take so much for granted.

In a book of this sort it is only possible to summarize very briefly the nub of legislative requirements, but I hope I have included sufficient information to give managers the basic information they need and safety advisers and students the trigger to remind them of more detailed requirements. One of the aims of the book has been to simplify the legislative requirements that are wrapped up in complex Parliamentary verbiage and it has done this by hacking away that verbiage to expose the basic objectives of health and safety laws in readable everyday English. I hope I have succeeded.

John Ridley
October 1997

Part 1 Law

The system of English justice was not designed but developed gradually over more than 1000 years. It is a system that has been used as a model in many parts of the world.

In parallel with the development of the system so the method of making laws has evolved and now follows a well proven procedure although there have been some radical changes in the past two or three decades.

Since 1972 a major influence on the subject matter of English laws has been the European Economic Community (now the European Union) where, as a Member State, the UK is required to incorporate into domestic laws the content of EU directives once they have been adopted by the Council of Ministers.

This part deals with the historical background to, procedures for making, administering and enforcing health and safety laws as well as covering the content of two of the main pieces of health and safety legislation.

1   Legal processes

The English legal system and its laws did not happen over night, they took thousands of years to develop. During that time they were adjusted and modified as the system of justice became more consistent across the nation. For the last 700 years there have been only relatively minor developments although the past two decades have seen a number of important changes in the role of members of the legal profession.

This section looks at the background to law in the UK, how laws are made and the effects of the increasing influence of the European Union on UK laws.

1.1 Historical background

After the Romans left England around AD 400, the country split into a number of small kingdoms each ruled by the local lord. This gave rise to great injustices and misuse by the lords of their powers. This situation continued until the latter part of the ninth century when the gradual unification of the country began. The English system of justice was not designed; it developed slowly over more than 1000 years through moves to correct particular problems and injustices, and it is still developing today. Some of the major steps in that development are summarized below.

Alfred the Great	871–901	• unified England • drew up a Book of Dooms or Laws • introduced an administrative structure for government
Edgar	959–975	• set up Shire, Hundred and Borough courts to administer the Dooms
Edward the Confessor	1042–1066	• first effective king • introduced King's Justices to administer more consistent justice across the country
William the Conqueror	1066–1087	• rationalized the existing legal arrangements • developed the English Constitution • established the 'Opposition' • established his own 'superior' courts to get away from the parochial nature of existing courts • initiated the doctrine of legal precedents
Stephen	1135–1154	• moved administration of the country away from the Crown • birth of the 'Civil Service'

| Henry II | 1154–1189 | • created King's Courts to try breaches of the King's Peace
• introduced trial by jury
• introduced the concept of Royal Writs to attract litigants to the King's Courts and give them fairer justice
• administered the courts through a Royal Council:
  – Court of Exchequer
  – King's Court or Bench
  – Court of Common Pleas
• appointed travelling (Circuit) judges to sit in the King's Courts around the country
• set up the framework of law that is followed today |
|---|---|---|
| John | 1199–1216 | • signed Magna Carta in 1215 giving many law-making powers to the lords and barons |
| Henry III | 1216–1272 | • established first Parliament of elected representatives of landowners and citizens |
| Edward I | 1272–1307 | • Edward the Law Giver
• made House of Lords the chief Appeal Court
• established three departments of the King's administration:
  – the Exchequer
  – the Chancery
  – the Wardrobe
• specialist professional legal groups became established, i.e. the Inns of Court: Clements, Lincolns, etc.
• passed statute declaring that English legal memory started when Richard I came to the throne in 1189 as a tribute to the efforts of Henry II in establishing a legal system |

From this time, with some small developments, the legal system that existed at Edward I's death has been followed to the present day.

However, in the last two decades or so there have been developments in the administrative arrangements within the legal system such as the setting up of Industrial Tribunals and the giving to solicitors the right to appear in the higher courts and to become judges.

1.2 Branches of law

English law divides into two major categories:

1 **Statute law** comprises those laws which are debated and made by Parliament. They are written and copies are available for all to purchase (from HMSO). Statute law is the law of the State and a breach of it is a criminal offence, hence it is sometimes referred to as criminal law.

2. **Common law** originally referred to a system of law common across the whole country. However, its meaning has changed to encompass **case law** – the body of law based on **judicial precedents**, i.e. the decisions of judges in earlier cases are binding in similar later cases. It is not *officially* written but is recorded in Law Reports.

Equity law supplements common law in that it deals with complaints against common law judgements where there may have been a distortion of justice, i.e. such as from undue influence by one party.

There is some overlap between these categories, for example, statute law can rely on interpretations from common law and common law cases can rely on breach of statutory duty to support a claim.

Statute law divides into:

1. **Public law** – dealing with matters involving the State and relating to the protection and well being of the public at large whether directly or indirectly. Typical of these laws are:

   - constitutional laws
   - employment laws
   - social security laws
   - criminal laws

2. **Private law** – statutory requirements that deal with matters regulating the relationship between 'private parties' or 'legal and natural bodies' such as in:

   - contracts
   - torts
   - property
   - trusts
   - succession

3. **Civil law** – a generic term covering all laws except criminal laws.

1.2.1 Legal actions

Criminal

- Law concerning the protection of the State, the community and the individual.
- Laws are passed or approved by Parliament.
- Written law, copies obtainable from HMSO.
- A breach is a crime or criminal offence.
- Criminal proceeding (prosecution) heard:
  – in Magistrate's Court for lesser cases
  – in Crown Court for more serious cases.
- Evidence must prove guilt 'beyond reasonable doubt'.

Civil

- Common law action.
- Between two or more private or corporate bodies.
- Based on precedent set by judgements in earlier cases.
- Proceeding (litigation) heard in County Court.

1.2.2 Interpretations

The meaning of words and phrases of statutes are often critical to the outcome of a case and may be decided by the judge when giving his judgement.

Decided cases – those cases in which, as part of a judgement, a word or phrase that is part of a law is defined, i.e. the meaning is **decided**. Once a meaning is decided, it applies to all laws. A good example is the phrase 'so far as is reasonably practicable' which was defined in a mining case in 1949.

Tort – a wrongful act or omission causing harm or damage to a person or body corporate and which is actionable in common law. Typical torts include:

- nuisance
- trespass
- negligence*
- breach of statutory duty*.

Rule of law

- No person or body is above the law
- There is one system of law for everybody.

Supremacy of Parliament

- There is nothing that Parliament cannot lawfully do.

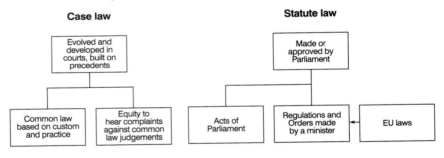

Figure 1.1 Sources of English law

1.3 Types of court

Matching the two types of law, i.e. statute and common, are two sets of court hierarchy, i.e. criminal and civil.

1.3.1 Criminal courts

(a) Magistrate's Court

- Dispenses summary justice, i.e. trial without a jury.
- Normally comprises a bench of between two and seven lay magistrates. Lay magistrates are not legally qualified and are unpaid. In court, they are advised on matters of law and procedure by the Clerk to the Court, who is a barrister or solicitor with at least five years experience.
- In some cities cases may be heard by a **stipendiary magistrate**, who sits alone but is a professionally qualified barrister or solicitor of at least seven years standing.
- Cases are heard in public.

* usual causes of actions in health and safety

Legal processes **7**

- Deals with minor criminal cases.
- Has limited powers of sentence:
  - maximum fine of £20 000
  - prison sentences up to three months for cases involving breaches of the conditions of a Prohibition Notice.
- Hears preliminary evidence in certain cases before committing them for trial in the Crown Court.

(b) *Crown Court*
- Court of **first instance**, i.e. 'hears' cases for the first time, the first level at which a case is officially recorded.
- Trial by jury.
- Hears trials on indictment.
- Both parties are represented by solicitor or barrister.
- Hears appeals from a Magistrate's Court.
- Will pass sentence on cases referred to it by a Magistrate's Court.

(c) *Court of Appeal*
- Hears appeals from the Crown Court.

(d) *House of Lords*
- Hears appeals from the Appeal Court.
- The final appeal court in the UK.

1.3.2
Civil courts

(a) *County Court*
- Covers a restricted geographical area.
- Hears a wide range of cases in the first instance.
- Presided over by a Circuit Judge or a District Judge who has limited powers of sentence.

(b) *Industrial Tribunal*
- Less formal than a court.
- Sits under a Chairman who is a solicitor.
- Chairman is supported by one representative appointed by an employer's organization and one appointed by an employee's organization.
- Procedure to be followed in making an appeal is laid down in *The Industrial Tribunals (Improvement and Prohibition Notices Appeals) Regulations 1974*.
- Applications to appeal to a Tribunal must be made within a prescribed period depending on the cause of complaint.
- Both appellant and defendant can represent themselves.
- Hears appeals against:
  - Improvement Notices
  - Prohibition Notices
  - refusal by an employer to give a safety representative time off with pay to carry out his functions or to attend training in his functions.
- Hears appeals on employment matters regarding:
  - redundancy
  - maternity leave
  - racial and sexual discrimination
  - unfair or constructive dismissal
  - equal pay.
- Appeals against a Tribunal's decision may only be made on points of law.

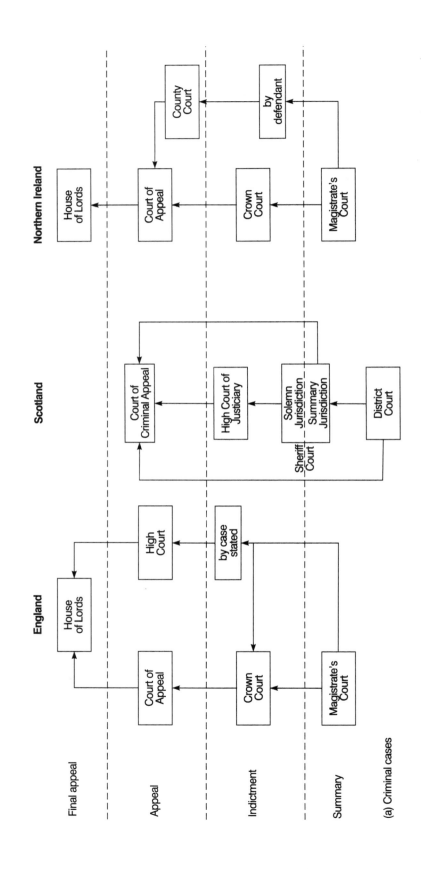

(a) Criminal cases

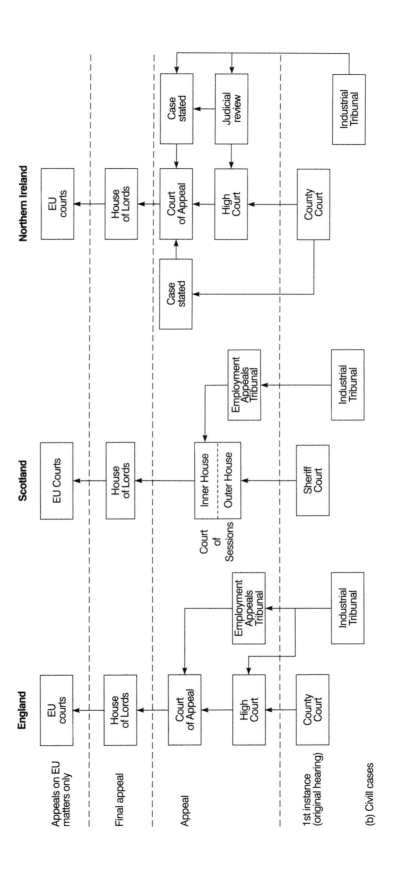

Figure 1.2 The different courts in the UK

(b) Civil cases

(c) Employment Appeal Tribunal	• Associated with the High Court. • Sits under a judge with 2 to 4 lay members all with equal say. • Hears appeals from an Industrial Tribunal on matters of law.

(d) High Court	• A group of three superior Courts having both first instance and appeal functions.
	• Divisions of the High Court are:
	  – Queen's Bench, dealing with criminal matters
	  – Family, dealing with adoption, marital disputes and property
	  – Chancery, dealing with companies, wills and bankruptcy.
	• Divisions of the High Court hear appeals from:
	  – Industrial Tribunals on matters of law
	  – County Courts.

(e) Court of Appeal	• Appeals heard by three judges.
	• Hears appeals from:
	  – Employment Appeal Tribunal on matters of law
	  – County Courts
	  – Divisional Court.

(f) House of Lords	• The ultimate point of appeal in the UK. • Appeals considered by three or more Law Lords.
(g) European Court of Justice	• Concerned only with matters relating to European Union legislation. • Cannot interfere with justice in a member state. • Judgements can require a member state to reconsider its own judgement.

Some of the courts and routes of appeal in Scotland and Northern Ireland are different from those in England. These are shown in Figure 1.2(a) and (b) for criminal and civil cases respectively.

1.4 Court procedures

Behaviour and procedures in court have developed over many hundreds of years and, in general, the procedure followed is very similar in all the courts, the main differences being the degree of formality and who is allowed to appear. Below are outlined the procedures followed in the first levels of court only, i.e. Magistrate's Court, County Court and Industrial Tribunal, since it is unlikely that many readers will become involved in the complex procedures of appeals which are handled by professionally trained legal people.

Described below are the procedures for English courts. Certain of the procedures followed in the courts in Scotland and in Northern Ireland differ from those in English courts. Some indications of these differences are given in Figure 1.2.

1.4.1 Magistrate's Court

A court of summary jurisdiction in which criminal cases, i.e. alleged breaches of, *inter alia*, health and safety laws, are heard by a 'bench' of between two and seven magistrates although in some cities the cases may be heard by a stipendiary magistrate sitting alone. Stipendiary magistrates are full-time, are paid a stipend and are usually barristers or solicitors with at least seven years legal experience.

Representation in court:

- The prosecutions in criminal cases are conducted by either a solicitor or barrister, except for health and safety cases, where the HSE (Health and Safety Executive) Inspector is authorized to conduct his own prosecution.
- The defendant can represent himself but in health and safety cases it is usual for the defendant to be represented by a solicitor or barrister. If there is a common law claim pending it is likely that the defendant's EL (Employer's Liability) insurers will insist on, and pay for, professional representation.

The procedure is as follows:

- The prosecuting inspector presents his own case.
- The defendant is normally represented by a solicitor or barrister.
- The charge is read out by the Clerk of the Court.
- The defendant is asked how he wishes to plead, guilty or not guilty.

If he pleads **guilty**:

- The prosecutor summarizes the circumstances of the alleged offence.
- The defence enters a plea for mitigation.
- Magistrates impose sentence.

If he pleads **not guilty**:

- The prosecutor presents his case.
- The prosecutor calls prosecution witnesses who give evidence under oath.
- The prosecutor questions them (examination-in-chief).
- The defending solicitor questions prosecution witnesses (cross-examination).
- The prosecutor sums up his case.
- The defending solicitor presents defence case.
- The defending solicitor calls defence witnesses who give evidence under oath.
- The defending solicitor questions witnesses (examination-in-chief).
- The prosecutor questions defence witnesses (cross-examination).
- The defending solicitor sums up the case for the defence.
- Magistrates retire to consider their verdict.
- Magistrates announce their verdict:
  – if not guilty, the case is dismissed
  – if guilty, a sentence is imposed.
- If a fine is imposed it must be paid by the accused – it cannot be paid by his insurers even though they pay the cost of the defence.

Magistrates have restricted powers of sentence – in most cases a fine not exceeding £5000 but in other serious cases, such as breaches of Prohibition Notices, they have power to impose a fine not exceeding £20 000 plus a custodial sentence of up to three months in jail.

If a fine is imposed, it is usual for the defence to ask for time to pay – otherwise the fine must be paid before they leave the court.

12 Legal processes

If the unsuccessful party is aggrieved by the verdict they can give notice of appeal, which if allowed, sets the sentence to one side until the appeal is heard.

A visit to a Magistrate's Court will give a good feel for the types of cases heard and for the procedures followed.

1.4.2 County Court

This court hears claims for damages arising out of an accident at work where the sum involved is less than £5000; above that sum the case is heard in the High Court.

Once the claim has been submitted and the defence entered, there is normally a great deal of pre-trial negotiation. This is carried out by the solicitors from each side and involves the gradual sifting and agreeing of evidence. In the case of a claim for damages the procedure is as follows:

- The plaintiff files a request to the court for damages and gives particulars of the claim.
    - The claim can be based on:
      negligence
      breach of statutory duty
      both.
- The court issues a summons on the defendant.
- Both sides are represented by solicitors or barristers.
- The defendant enters a pleading refuting the claim.
- Contact is made between the solicitors for both sides regarding common evidence, exchanging information and endeavouring to reach a settlement.
- If agreement is not reached there may be a pre-trial review (interlocutory hearing) to clarify the issues, agree evidence to be used and attempt to find a settlement.
- The case goes to trial and follows a similar court procedure to that used in the Magistrate's Court.
- If an appeal is made it is directed to the Divisional Court or to the Court of Appeal.

The payment of any damages agreed or awarded is made by the employer's EL insurer who also bears the costs of the case.

'You expect me to believe that?'

1.4.3 Industrial Tribunal

This hears cases arising from breaches of employment law where an employee complains against a decision of the employer in situations covered by the legislation. It also hears appeals against Improvement and Prohibition Notices issued by an enforcing officer. The tribunal is a civil court because the matters it considers relate to a difference of opinion between the employer and the employee, or the employer and the individual enforcing officer, over the interpretation of the law and not to a breach of the law.

The procedure in taking a complaint to an Industrial Tribunal is as follows:

- The appellant sends notice of appeal in writing to the secretary of the tribunal.
- An appeal must be made within 21 days if it relates to an Improvement or Prohibiton Notice.
- The respondent is sent a copy of the complaint.
- A date is set for the hearing.
- The tribunal consists of a legally qualified chairman supported by two lay members, one nominated from the employer's organization and one from the employee's organization.
- Both parties can present their own cases or can be represented (representatives need not be legally qualified but can be a friend such as a trade union official).
- The proceedings are under oath but much less formal than a court.
- Each side presents its case.
- The tribunal can ask questions.
- The tribunal makes its decision known, which is binding on both parties.
- No costs are awarded.
- Any appeal is made to the Employment Appeal Tribunal.

All court cases are time consuming (and hence expensive) for the employer, especially so in an Industrial Tribunal case where the employer is so much more closely involved.

1.5 Law-making in the UK

The process for making laws in the UK varies slightly between the statutes and the subordinate legislation, i.e. Regulations and Orders. All are in written form and copies may be obtained from HMSO.

1.5.1 Statutes

The process for making statute law is as follows:

- The originator receives Parliament's permission to submit a proposal. The originator could be one of the major political parties or an MP (Private Member's Bill).
- A Green Paper outlining the aim of the proposal is prepared for discussion.
- A White Paper giving the policy and intent of the proposal is prepared.
- A Bill, which contains the proposed wording of the new statute, is presented to Parliament for its consideration.
- Parliament:
  (1) gives it a first reading
  (2) gives it a second reading
  (3) submits it to Committee for detailed examination of its content and implications
  (4) gives it a third reading.

- The Bill goes to the 'other house' (House of Lords) where a similar procedure is followed.
- When both Houses have agreed the content, the Bill is given a final reading before being submitted to the Sovereign for the Royal Assent which, by tradition, is never refused.
- On receiving the Royal Assent it becomes an Act and is entered in the Book of Statutes and becomes part of the statute law of the country.

Within an Act may be terms that empower a Minister of the Crown to make subordinate laws to supplement the main Act.

A subordinate law may be:

(1) An Order:

- Usually concerning an administrative matter such as the timing of the implementation of particular sections of the main Act.
- Prepared with minimal consulation.
- Approved by Parliament by negative vote, i.e. no objections.

(2) A Regulation:

- Prepared by the Health and Safety Commission/Executive (HSC/E) on behalf of the minister.
- Statutory requirement for interested bodies to be consulted.
- A Green Paper is prepared for limited discussion.
- A White Paper is prepared for limited discussion with interested bodies such as the CBI, TUC and affected industrial sector representatives.
- A Consultative Document is issued setting out the proposed wording, giving an explanation of the reasoning behind the proposal and calling for comments.
- Once comments are received, a draft regulation is prepared which may be subject to limited consultation with CBI, TUC and particular industry.
- A final proposal is submitted to the HSC for transmission to the minister.
- The minister lays the proposal before Parliament.
- It is approved by Parliament using either:
  - **negative voting procedure**, i.e. it lies on the table in the House for 30 days and if there are no objections it is automatically passed
  - **positive voting procedure**, i.e. it lies on the table of the House for 30 days during which time any MP can call for a debate on it. At the end of 30 days a positive vote is necessary to pass it.
- It then becomes a *statutory instrument* with the full power of law.

Acts and Regulations normally have a run-in period before full implementation. This allows HMSO time to print copies and industry to adapt to the new requirements. In some cases, implementation may be phased over a number of months or years.

1.6 Law-making in the EU

In 1972 when the UK joined the then European Economic Community (EEC) now the European Union (EU), the UK government agreed to be bound by the various laws adopted by the Council of the European Communities. EU laws are 'adopted' by Council, not 'passed', and once adopted, they have immediate applicability on the government and its employees but do not apply to the private sector until national laws, incorporating the content of the directive, have been passed by Parliament.

Originally the Treaty of Rome required unanimous assent for a matter to be adopted, but in 1986 the *Single European Act* was adopted, modifying the procedure. The Act did the following:

- Set 31.12.1992 as the date by which the single Internal Market must be established.
- Brought in *qualified majority voting** on issues aimed at the establishment and functioning of the Internal Market, including health and safety (art. 100A).
- Called for harmonization of working conditions across Member States (art. 118A).
- Established new 'co-operation' procedure for adopting legislation (see Figure 1.3 below).
- Required directives not to prejudice the development of small and medium enterprises (SMEs).
- Established a *Court of First Instance* allowing direct access to EU courts by individuals and companies on matters concerning EU laws.

EU legislation comprises:

- Regulations
  - have direct applicability in the Member States at which they are aimed
  - apply mainly to the iron and coal industries and are rarely used for health and safety matters.
- Directives
  - the main body of EU law, used extensively in all areas of Community activities.
- Decisions
  - usually about a specific subject and may be aimed at a particular Member State where they have direct applicability.

In drafting health and safety directives there is wide consultation with interested sectors, both public and private. The European Parliament has a considerable say.

The drafting procedure follows the 'new approach to legislative harmonisation' whereby the directive itself contains broad objectives to be met and gives in annexes details of particular areas requiring attention, relying on *harmonized standards* (EN standards) to provide detailed requirements for conformity with the directive.

Figure 1.3 shows diagrammatically the process for promulgating a directive.

---

* A system whereby each Member State is given a number of votes depending on its size. 70% of votes in favour are required for adoption.

16 Legal processes

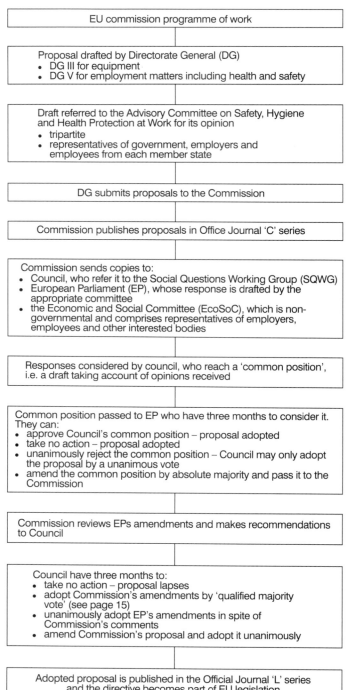

Figure 1.3 Procedure leading to adoption of a directive

1.6.1
European harmonized standards

European standard-making bodies:

- Comité Européen de Normalisation (CEN), for mechanical standards.
- Comité Européen de Normalisation Electrotechnique (CENELEC), for electrical standards.
- Common Central Secretariat.

Their work is overseen by General Assembly and Technical Boards of representatives of the standard-making bodies of Member States. Member States include EU members, European Free Trade Association (EFTA) members plus affiliates from Eastern European and Middle Eastern countries.

The leading UK body is the British Standards Institution.

CEN and CENELEC are sponsored by the EU. They work closely with their international opposite numbers, the International Standards Organisation (ISO) and the International Electrotechnical Commission (IEC), respectively.

Harmonized standards take precedence over national standards and are recognized in the UK by a prefix to the standard number, i.e. BS EN.

Equipment conforming to harmonized standards and complying with the *Supply of Machinery (Safety) Regulations 1992* can be marketed throughout the EU without restriction.

Harmonized standards are divided into four categories:

1 Type A, relating to fundamental safety concepts and principles

   - general safety principles
   - instruction handbooks
   - rules for drafting standards, etc.

2 Type B1 for safety aspects applicable to a large number of machines

   - safety distances
   - hand and arm speeds
   - noise and vibrations
   - hydraulic/pneumatic control systems
   - safety symbols, etc.

3 Type B2 for safety related devices that may be used on a variety of machines

   - two-hand controls
   - electro-sensitive safety systems
   - pressure-sensitive mats
   - interlocking devices, etc.

4 Type C for devices that are specific to certain types of machinery

   - cold forming of metal
   - industrial robots
   - mechanical handling equipment
   - construction equipment, etc.

2 Health and safety laws

The early health and safety laws, the *Factories Act 1961* (FA) and the *Offices, Shops and Railway Premises Act 1963* (OSRP), were aimed at correcting identified wrongs and tended to be very prescriptive in their content (i.e. they laid down how the wrong was to be corrected) and restrictive in their field by applying only to the particular operation, process or premises defined in the Act. A number of regulations made under these Acts are still in effect.

Health and safety laws have, in the past, evolved to protect against the hazards brought about by developing technology. While this still applies today, a greater emphasis is now put on the part that the employer can play. Attitudes have changed from protecting machinery to protecting people and are progressively looking at anticipating hazards (risk assessments) rather than waiting for them to manifest themselves through accidents.

Current health and safety legislation revolves round the *Health and Safety at Work, etc. Act 1974* (HSW), which is proscriptive (i.e. it sets out the objectives to be achieved without specifying how). Regulations made under HSW apply to all employments unless the regulation itself restricts its application.

HSW places basic obligations on the employer, the employee, tenants in multi-occupancy premises and on the landlord. HSW contains conditions that allow the making by the Minister of subordinate laws – statutory instruments known more commonly as Regulations and Orders – to enable particular requirements to be placed on a whole gambit of employment and workplace situations. These subordinate laws are gradually building up to provide a great body of health and safety legislation that is superseding and updating the existing well-established if outdated laws, and making them more relevant to modern-day employment and processes.

The majority of health and safety regulations concerning employment at work, including those that emanate from the EU, are made by the Department of the Environment (DoE) (via the Health and Safety Commission/Executive [HSC/E]) under powers contained in HSW.

To stimulate the growth of free trade in the EU a number of directives have been adopted that lay down safety requirements for goods and equipment. The drawing up of regulations to incorporate the content of these particular directives into UK laws is carried out by the Department of Trade and Industry (DTI) under powers contained in the *European Communities Act 1972*.

2.1 A historical perspective on health and safety laws

This section covers very briefly some of the salient stages in the development of health and safety legislation in the UK.

1556	Part of a book on metal mining by Dr Agricola dealt with diseases of miners
1567	Treatise on diseases in mining and smelting by Dr Paracelsus
1690	*Boson* v. *Sandford* – first judgement that established the doctrine of vicarious liability of the employer, i.e. the employer being responsible for what his employees do while at work
1700	Book on trade diseases published by Italian physician Bernadino Ramazzini
late 1700s	Industrial Revolution
1784	Fever epidemic in Lancashire claimed many lives, particularly of young children
1795	Manchester Board of Health set up
1800	Combination Acts outlawing trade unions
1802	Act for the preservation of the health and morals of apprentices and others in cotton mills (the first health and safety Act)
1819	Further Act, prohibiting in mills the employment of children under nine years of age, reducing the hours of work for 9–16 year olds to twelve per day and prohibiting night work for under 16s
1824	Repeal of the Combination Acts
1832	Leeds doctor Charles Turner Thackrah published first English book on occupational diseases
1833	Act further reducing the hours of work of women and young persons, also authorizing the appointment of four Inspectors of Factories
1840	Et seq. saw the development of the doctrine of Breach of Statutory Duty as grounds for a claim in common law
1842	First Act relating to coal mines, banning the underground employment of women and children
1844	Further Act dealing with hours of work; also contained first requirement regarding machinery safety by prohibiting the cleaning of moving machines – but still applicable only to mills

Early industrial injury

1864	Safety legislation extended beyond mills to include some manufacturing factories
1871–5	850 boiler explosions resulting in death or serious injury
1878	Factory and Workshop Act – extending legal protection to nearly all manufacturing industries; first Act to consolidate safety legislation
1881	Boiler Explosions Act – laying down safety measures to be incorporated into boilers
1897	Workmen's Compensation Act – applicable to limited range of industries only
1901	Factory and Workshop Act – a further consolidating Act including, for the first time, powers for the Minister to make Regulations
1906	Workmen's Compensation Act – extended to all wage earners within certain wage limits
1937	A further consolidating Factories Act
1961	The last consolidating Factories Act
1963	The Offices, Shops and Railway Premises Act
1972	UK became a Member State of the European Economic Community and subject to EEC directives
1974	The Health and Safety at Work, etc. Act
1986	The Single European Act
1992	A series of Regulations incorporating the requirements of EEC directives into UK law, repealing considerable portions of the Factories Act 1961, the Offices, Shops and Railway Premises Act 1963 and revoking some associated Regulations
31st December 1992	The Single European Market established

Health and safety laws continue to be made at an accelerating rate, many of them as a result of our membership of the EU.

2.2 The structure of health and safety laws

Legislation is structured in the following way:

Primary legislation

- Acts
  - known as statutes
  - fully debated in both Houses of Parliament
  - become law when they receive the Royal Assent.

Secondary legislation (sometimes referred to as subordinate legislation)

- Regulations
  - known as statutory instruments
  - drafting power delegated to a Minister of the Crown
  - interested bodies must be consulted
  - voted on by Parliament which gives them the power of law.

- Orders
  - known as statutory instruments
  - prepared by the Minister concerned and issued by him
  - deal with administrative matters such as implementation of an Act.
- Bylaws
  - made by local authorities under powers given to them by the Act which established them
  - they apply only within the local authority area
  - are not part of criminal law
  - can be challenged in court.

The following supporting documents can be used in the legislative process:

- approved Codes of Practice
  - have a quasi-legal status
  - can be used in court as evidence of acceptable levels of compliance
- various health and safety guidance booklets published by the Health and Safety Commission and Executive
- British and European harmonized standards
- industry-based standards often with HSE input.

Prior to 1974 the main health and safety Acts were the *Factories Act 1961* and the *Offices, Shops and Railway Premises Act 1963*. Both these Acts applied only in the premises defined in them. Similarly, any subordinate legislation made under either of them applied only in the premises covered by the main Act. However, legislation made since the coming into effect of the *Health and Safety at Work, etc. Act 1974* now covers all employment except domestic service.

Currently the content of EU Directives are incorporated into UK law by means of Regulations. Those concerned with employment, i.e. the use of machinery and equipment, are drafted by the HSC and made under HSW. Those concerned with the standard of safety of goods are made by the Department of Trade and Industry under the *European Communities Act 1972*.

2.3 The Health and Safety at Work, etc. Act 1974

The Health and Safety at Work, etc. Act 1974 (HSW) is based on the recommendations of the Robens Report and sets out in principle the objectives to be achieved for ensuring high standards of health and safety in the workplace. This Act is gradually superseding the earlier *Factories Act* and the *Offices, Shops and Railway Premises Act*. It is being reinforced by regulations made under its delegating powers.

A great deal of emphasis is placed on the role and responsibilities of the employer as the person who decides and controls what goes on in the workplace, but it does not forget that employees also have a contribution to make.

The general requirements for achieving high standards of health and safety at work are contained in ss. 1–9 of HSW, laying down broad objectives to be met and placing duties for meeting those objectives on the employer, the landlord, tenants, employees and suppliers.

The Act itself is divided into a number of parts:

Part 1
- Starts with a statement of the intent of the Act.
- Outlines the general duties placed on employers, landlords, tenants, employees and suppliers.
- Authorizes the setting up of the Health and Safety Commission and Executive and lays down how they should operate.
- Outlines the procedures to be followed in making regulations and issuing approved codes of practice.
- Lays down the enforcement powers of inspectors.
- Sets the criteria for obtaining and disclosing information that inspectors may acquire in the course of their investigations.
- Lists the offences under this Act and types of legal proceedings.
- Outlines appeals against action taken by an inspector.
- Denies the right of civil action based on breach of statutory duties placed by this Act.

Part 2
- Allows for the continuation of the Employment Medical Advisory Service.

Part 3
- Once dealt with building standards but has been repealed, its subject matter being incorporated into the *Building Act 1984*.

Part 4
- Deals with various administrative matters and interpretations.

Schedules
- Lists the circumstances and subject matters on which Regulations can be made (the list is pretty comprehensive and covers virtually all work situations).

More details of particular duties placed by this Act are dealt with in Part 2 of this book.

2.4 Health and safety law enforcement

The general responsibility for enforcing health and safety laws lies with the Health and Safety Commission (HSC) but the actual enforcement is carried out by the Health and Safety Executive (HSE). However, because of the very wide range of areas to be covered, the HSE has delegated certain aspects or areas of enforcement to other bodies.

2.4.1 Enforcement responsibilities

- HSE – factories, manufacturing premises, fire prevention (where the high fire risk is in the process), railways, offshore industries, nuclear, agriculture, mines and quarries
- Local authorities – offices, shops, warehouses and non-machinery premises

Health and safety laws **23**

- Fire Authority – fire prevention except where the high fire risk is in the process, means of escape, issue of Fire Certificates.

2.4.2 Authority of Inspectors

The work of enforcement is by inspectors:

- whose powers derive from ss. 19 and 20 of HSW
- who get their authority from a warrant they must carry when inspecting
- the warrant must be signed by a senior officer of the authorizing body.

2.4.3 Powers of Inspectors

HSE inspectors and local authority inspectors (Environmental Health Officers [EHO]) have the following general powers:

- Entry to any premises at any time work is being carried out.
- To obtain information:
  - take measurements and photographs
  - inspect documents but not 'privileged' documents
  - require persons to give information relevant to their investigations (failure to do so is an offence).
- Issue Improvement Notices.
- Issue Prohibition Notices.
- Not to pass to a third party information from investigations (to do so is an indictable criminal offence).
- Prosecute in a Magistrate's Court for alleged breaches of health and safety laws.

In exercising these powers they do the following:

- Carry out general inspections of workplaces (usually the more hazardous the workplace the more frequent the inspections).

An inspection

24 Health and safety laws

- Investigate accidents at work.
- Investigate complaints about working conditions (this is given a high priority in their work).
- Give advice on specific health and safety problems.
- Pass to representatives of the employees factual information on what they have found.

Fire officers have broadly similar powers except that they do not have authority to prosecute and their inspections are restricted to Fire Certificate matters.

2.4.4 Organization for enforcement

Notes on HSE enforcement organization (shown diagrammatically on p. 25):

1 Within the Regions are point of co-ordination of safety standards for particular industry sectors known as Sector Groups.
2 To advise it on matters for improving health and safety within an industry, the HSC has established a series of Industry Advisory Committees (IAC). These IACs are tripartite with the Sector Groups providing the secretariat.
3 Local authority inspectors, the Environmental Health Officers, have similar powers to an HSE inspector but report to the local authority. They may seek advice from the HSE on particular matters.

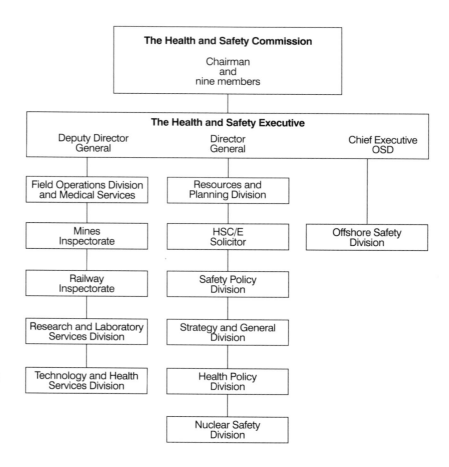

Figure 2.1 The central administrative organization of the Health and Safety Commission and Executive (HSC/E)

Health and safety laws **25**

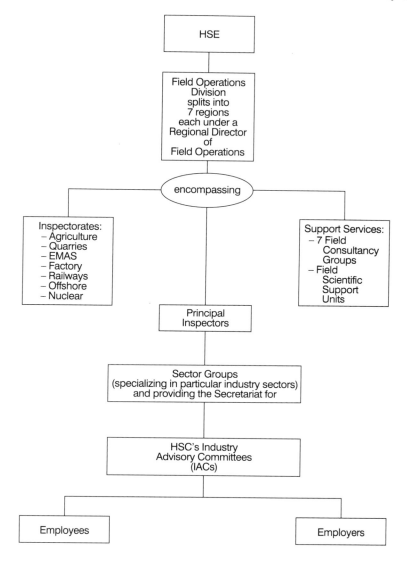

Figure 2.2 Diagram of HSE enforcement organization

4 Fire Prevention Officers report through the Fire Authority to the Home Office but work closely with the HSE on industrial matters.

2.5 The Management of Health and Safety at Work Regulations 1992

These regulations apply to all employment and supplement the requirements of HSW and other extant legislation. Where there is overlap, the more stringent requirements take precedence.

They recognize that the major responsibility for health and safety at work lies with the employer but that the employee has a contribution to make (r.12). They place particular responsibilities on those who provide *contract, short-term* or *temporary* workers (i.e. the employment agencies) even though the control of how the work is carried out is not in their hands.

The main contents of the Regulations are:

r.3 Employers are required to:
- carry out a **risk assessment** of any work activities that may put anyone at risk
- take particular account of:
  - the inexperience of young persons
  - risks to women who are of child-bearing age, pregnant or nursing mothers, from working methods and the use of chemicals
- use the assessment to determine the precautions necessary to protect against the identified hazards. The precautionary measures to overcome or reduce the hazards identified will depend on the work circumstances
- repeat the assessment if the work changes
- record the assessment if there are more than five employees. [See Section 3.4 on risk assessment]

r.4 Employers are required to:
- integrate health and safety into all management systems
- make health and safety an inherent part of all operations and work activities
- employ a systematic approach so that the provision and use of protective and precautionary measures are properly planned, organized, executed, maintained and reviewed.

r.5 Health surveillance is to be provided to employees:
- where an occupational disease or health risk is identified
- if an identifiable disease or health risk is likely under normal foreseeable work conditions
- carried out by a suitably qualified person, i.e. occupational health nurse or doctor.

r.6 Every employer must have:
- access to competent health and safety advice
- this service can be by a suitably qualified employee or as a bought-in service
- such advisers must be given time to fulfill their health and safety function
- where non-employees (consultants) are used:
  - they must be given information on any known or suspect hazards in the workplace and any special conditions that apply to the work
  - the qualifications of such consultants are not specified but employers must satisfy themselves as to the consultant's competence. (A Registered Safety Practioner (RSP) or full corporate member of IOSH would be evidence of competency as would successful completion of the NEBOSH Diploma examination supported by the appropriate NVQ or SVQ at level 3 or 4 in Occupational Health and Safety Practice.)

r.7 Where danger exists in the workplace:

- those exposed must be told about it
- they must be told about the arrangements in place for their protection and how to avoid the danger
- employees must be trained in what to do in the event of imminent danger
- there must be suitably trained staff to ensure the emergency procedures are followed – such as fire marshal or a 'responsible person' appointed to deal with injury emergencies.

r.8 All employees are to be provided with 'comprehensible* and relevant' information on:

- the risks they face in their work including any that result from multi-occupancy of the premises
- precautions and preventative measures in place to protect them
- emergency procedures and who the emergency marshals are
- where a child is employed the parents must be given the above information.

r.9 Where two or more employers share the same premises (multi-occupancy):

- they must co-operate in meeting statutory obligations
- they must keep each other informed of specific risks arising from their particular operations
- these obligations may be met by appointing a safety co-ordinator for the premises
- where self-employed persons share premises they too must co-operate for the common good
- where someone other than the employer has control over part of the work premises (i.e. a landlord), he too must co-operate with his tenant employers to ensure health and safety in the various workplaces. This is in addition to any responsibility he carries for common areas under his direct control, i.e. entrance halls, stairs, lifts, etc.

r.10 Where non-employees carry out work whether as contractors, servicemen, or 'temps' from an agency, the client or person who has control of the area where they are to work must inform their employers, or the agency who sent them:

- of the work that is to be carried out
- of any risks associated with that work
- of the precautions in place to protect them.

r.11 In allocating work the employer must:

- take account of the employee's abilities, both mental and physical, to do the work required of him

---

* allowing for those with English language difficulties.

- ensure employees are adequately trained:
  - on joining the company
  - on exposure to new risks because of:
    * moving to a different job
    * the introduction of new techniques or materials
    * changes to existing equipment or the introduction of new equipment
    * the introduction of new systems of work
- ensure that the training:
  - is on-going
  - is repeated periodically (refresher training)
  - takes account of changing risks
  - takes place in working hours (or be paid for as overtime).

r.12 Employees must:

- follow the training they have been given when using any equipment, substance or safety device provided by the employer
- follow the employer's instructions to ensure legal requirements are met (see also s.7 HSW)
- report to their employer any serious risks to health or safety they may find
- report any shortcomings in the safety arrangements.

r.13 Where temporary workers are employed, i.e. those on fixed-term contracts, they must be provided with information on:

- any special qualifications or skills necessary to do the work safely
- any health surveillance associated with the work.

Where temps from an agency are used, the agency must be provided with information on:

- any special qualifications or skills necessary to do the job safely
- any features of the job likely to affect the health and safety of the temporary workers
- The agency must pass this information on to the workers concerned and the employer must check before they start work that the agency has indeed passed the information on.

Where young persons are employed, the employer must:

- ensure that the young person is not put at risk because of:
  - inexperience
  - lack of awareness of dangers
  - physical immaturity
- not employ them:
  - where the work is beyond physical and psychological ability
  - on processes likely to affect their health

> – where subjected to radiations
> – where experience is necessary to recognize danger
> – in extremes of heat or cold
> – in noisy areas
> – where exposed to vibrations.
>
> Where women who are of child-bearing age, pregnant or nursing mothers, are employed they must:
>
> - not be employed in work that can put their child's health at risk
> - be moved
>   – to safer work or
>   – have their hours of work reduced or
>   – be suspended from employment.
>
> r.15 These Regulations cannot be used as the basis for a civil action except where the action arises from the employment of:
>
> - a young person
> - a woman of child-bearing age
> - a pregnant woman
> - a nursing mother.

While these Regulations add considerably to the obligations placed on employers by HSW and other Regulations made under it, they require no more than a caring employer would provide.

2.6 Other current health and safety legislation

A number of Acts not having direct relevance to health and safety, and not referred to in the above or the following text, nevertheless contain features that effect health and safety issues:

- *Environmental Protection Act 1990* deals with discharges to the atmosphere, to water courses and the disposal of waste on land. Some of the substances involved can pose a risk to the health of the community and to the ecology of the area.
- Various of the complex laws concerning employment which give protection to employees are relevant. In particular, s. 28 of the *Trade Union Reform and Employment Rights Act 1993* which gives employees protection against dismissal where health and safety is an issue.
- *Sex Discrimination Act 1975* requires that women be given equal employment opportunities to men and that they are not discriminated against because of their sex except where to do so would cause a breach of another statutory requirement, such as handling certain chemicals that could put a woman's childbearing ability at risk.
- *Race Relations Act 1976* requires that there be equal treatment of all people regardless of race or ethnic origin. However, again exceptions can be made where a breach of other legislation may be caused, such as Sikhs not wearing hard hats on a construction site or having beards in certain food processes.

Further Acts that have health and safety connotations include:

- *The Consumer Protection Act 1987*
- *The Employer's Liability (Compulsory Insurance) Act 1969*
- *The National Insurance (Industrial Injuries) Act 1946*
- *The Occupier's Liabilitiy Act 1984*
- *The Offshore Safety Act 1992*
- *The Disability Discrimination Act 1995*

together with a developing range of associated Regulations.

Part 2 Management

The management of an organization is a complex mix of people and systems covering a very wide range of activities and functions. The function of management is to draw all these aspects together into a coherent whole and guide it towards the organization's goal. In health and safety, while personalities inevitably play a part, there are a number of established and documented facets of the part that management (or more truly the manager) can play in ensuring employees go home in the same healthy state in which they arrived for work.

This part looks at some of the management techniques that are important in achieving high standards of health and safety in the workplace.

3 Management responsibilities

While many managers may be remote from the work areas, they exert a great influence, both formally and informally, through 'setting the safety tone' for the organization. Attitudes in the boardroom are manifest on the shopfloor even though there is no direct communication link. Thus if the Board is concerned about health and safety this will show in a high level of safety performance in the organization as a whole.

Thus, in fulfilling their role in the enterprise, managers are responsible for ensuring that the part of the organization within their control is operating at maximum efficiency, not only in production and quality but also in health and safety.

This section looks at the health and safety responsibilities placed on managers and some of the techniques they can use to meet those responsibilities.

3.1 The role of management

Managers have a direct bearing on health and safety since they have control and can give instructions. Managers are also the focal point of a lot of employee attention and the manner of their behaviour (their example) and their 'seen' concern for health and safety matters is an important factor in employee attitudes.

Managers can influence safety performance by:

- setting policies that require high safety performance
- providing resources to achieve the aims of those policies
- ensuring that the resources provided are used properly and effectively
- giving local managers sufficient freedom and authority to achieve high standards of health and safety in their own way (encourage their initiative and commitment)
- holding local managers accountable for their safety performance
- demonstrating a commitment to safety by:
  - personal involvement in health and safety matters
  - encouraging high standards of safety by a proactive approach
  - ensuring health and safety matters are included on board agendas
  - giving health and safety equal consideration with production, finance and sales, etc.

34 Management responsibilities

... a proactive approach to safety?

— being knowledgeable in health and safety issues when visiting the working areas and discussing them with employees.

In health and safety matters, the role of the employer and manager is to lead and control. 'How' is dealt with in the following sections.

3.2 Responsibilities for health and safety

UK laws put obligations on employers (i.e. the managing director, chief executive, partner or proprietor) to ensure the health and safety of employees. In large organizations the day-to-day performance of these obligations is delegated to subordinate managers who are given suitable authority to meet them.

Note that health and safety *responsiblities* of the employer cannot be delegated, it is only the *performance* or carrying out of those responsibilities that can be delegated.

The obligations placed on the employer by:

- the *Health and Safety at Work, etc. Act 1974* (HSW)
- the *Management of Health and Safety at Work Regulations 1992* (MHSW Regs)

- the *Workplace (Health, Safety and Welfare) Regulations 1992* (WHSW Regs)

are qualified by the phrase 'so far as is reasonably practicable'. The meaning of this phrase was decided in 1949 by Lord Justice Asquith in *Edwards* v. *National Coal Board* [1949] 1 All ER 743 as:

> 'Reasonably practicable' is a narrower term than 'physically possible' and implies that a computation must be made in which the quantum of risk *is placed in one scale and the* **sacrifice**, *whether in time, money or trouble, involved in the measures necessary to avert the risk is placed in the other; and that, if it be shown there is a gross disproportion between them, the risk being insignificant in relation to the sacrifice, the person upon whom the duty is laid discharges the burden of proving that compliance was not reasonably practicable. This computation falls to be made at a point in time anterior to the happening of the incident complained of.*

Broadly interpreted this means that protective measures should be considered before any accident happens, and that if the cost of those protective measures is excessive for the benefits derived, the provision of the measures is not reasonably practicable. However, if this argument is used as a reason for not providing protective measures it may be necessary to justify it in court should an accident occur or should an inspector query it.

The obligations placed on employers by s. 2 of HSW are *so far as is reasonably practicable* to:

- Make arrangements to ensure the health and safety of employees.
- Provide plant and equipment that is safe.
- Implement systems of work that are safe.
- Ensure the safe use, handling, storage and transport of both articles (equipment) and substances (chemicals).
- Keep employees and others (contractors, visitors, etc.) on the site informed on health and safety matters and arrangements.
- Provide adequate health and safety instruction and training.
- Ensure supervision is adequate and competent.
- Keep the workplace in good condition.
- Ensure the work environment does not put anyone's health at risk.
- Provide suitable welfare facilities.
- Have a written safety policy if more than five employees.
- If unionized, to recognize union-appointed safety representatives.
- Consult with safety representatives and employees on health and safety matters.
- Establish a safety committee when requested by two or more safety representatives (but there is nothing to prevent a voluntary safety committee being set up).
- In shared premises, to co-operate with neighbours on health and safety matters.
- Not emit noxious or offensive fumes or dusts.
- Not charge for PPE (personal protective equipment).

Duties are also placed on:

- employees:
  - to take care of themselves and others who may be affected by their acts or omissions
  - to co-operate with the employer in complying with statutory requirements
- everyone:
  - not to interfere with or misuse anything provided to meet a statutory requirement
  - report any dangerous situation to the local manager
- landlords:
  - to maintain common areas of premises under their control in a safe condition, i.e. entrances, gangways, stairs, lifts, etc.
  - to ensure any plant provided is safe to use
- suppliers of both substances and equipment:
  - carry out suitable tests to ensure equipment is safe for use
  - carry out tests to determine the chemical characteristic of a substance
  - provide the user with:
    * information on the limitation of its design
    * written instruction for its safe use
    * details of hazards of substances and precautions to be taken (safety data sheets).

Further obligations are placed on employers by MHSW and WHSW Regulations to:

- carry out risk assessments
- integrate health and safety into management systems
- provide health surveillance where the risk warrants it
- have available qualified health and safety adviser(s).

Where other more specific requirements arise those obligations are dealt with in the relevant part of the following text.

3.3 Safety policy

A duty is placed on employers (those with more than five employees) by s. 2 (3) of the HSW to:

> ... prepare and as often as may be appropriate revise a written statement of his general policy with respect to the health and safety at work of his employees and the organisation and arrangements for the time being in force for carrying out that policy ...

Thus an employer must have in writing not only the policy statement but also details of the organization he has established for implementing that

Management responsibilities **37**

policy plus information on the rules and procedures by which the aims of the policy are be achieved.

Hence, meeting these obligations can be considered in three parts:

1   The **safety policy** should:

    - state the aims of the organization for ensuring the health and safety of those who work in it or may be affected by its activities, i.e. contractors, visitors, neighbours, members of the public, etc.
    - refer to measures that exist to consult with employees on safety matters
    - indicate sources of expert safety advice
    - refer to the means for disseminating health and safety information
    - mention the importance of the part that the employees can play in achieving safe working conditions
    - be:
      – in writing
      – signed by the head of the organization
      – dated
      – brought to the notice of all employees
      – monitored
      – reviewed periodically
      – re-issued as necessary.

2   The **organization** for implementing the policy should include:

    - the name of the director with overall responsibility for health and safety
    - the names of other members of the organization with safety responsibilities
    - the safety responsibilities held by each member
    - the responsibilities of subordinate managers for preparing health and safety policies for their departments

Encouraging employee co-operation

- relationships with recognized or other unions
- routes for joint consultation on health and safety matters
- specialized responsibilities for safety advice, training, monitoring the policy, etc.

3 The **arrangements** for achieving the aims of the policy should include:

- a list of the agreed safety rules and procedures with a brief summary of each
- existing safe systems of work
- provisions for safe maintenance
- procedure for carrying out risk assessments
- accident reporting and investigating procedures
- controls for the safe use of chemicals
- measures for the safe introduction of new machinery and chemicals
- arrangements for dealing with emergencies including evacuation
- methods for the dissemination of information
- training facilities
- procedures for joint consultation including safety committee meetings
- issue and use of personal protective equipment
- steps taken to protect the environment
- welfare arrangements and facilities
- any other safety related matter particular to the organization.

For the safety policy, with its supporting arrangements, to be effective there must be means in place for checking that the agreed procedures and methods are being followed, that they are effective, and for implementing any changes those checks may show to be necessary.

3.4 Risk assessments

The term 'risk assessment' comes from the insurance industry and was one stage in their process of determining and spreading the liabilities they carried. It has been adopted into health and safety and its meaning widened to cover a spectrum of activities from the initial identification of a hazard to the establishment of safe working conditions.

It has become a central feature of new health and safety legislation and, while it was required by some earlier legislation, virtually all new legislation that is concerned in any way with matters that affect the health, safety and welfare of people at work does, and future legislation will, include a requirement to carry out a risk assessment.

Effectively, a **risk assessment** is a means whereby an employer can manage properly the risks faced by his employees and ensure their health and safety is not put at risk while at work.

The Management Regulations place specific responsibilities on employers to:

- Identify hazards that pose risks to the health and safety of employees.
- Carry out 'suitable and sufficient' risk assessments of hazards identified.

- Decide what is 'suitable and sufficient' in the light of their operating circumstances.
- Cover in the assessment:
  - all equipment, both existing and new
  - materials and substances.
- Give priority to protecting whole work force rather than individuals.
- Consider any risks from their operations that may affect non-employees such as agency and contract workers, contractors, visitors and those with a right of entry such as postmen, utilities employees, delivery drivers, etc.
- Appoint an assessor:
  - to carry out assessments
  - who has knowledge of:
    * the work processes
    * health and safety legislation
    * current health and safety standards for the industry.
- Give the assessor time to carry out the assessments during working hours. (The assessor could be a supervisor or chargehand who has had health and safety training.)
- If more than five employees, record* the results of risk assessment.

3.4.1 Definitions

Certain terms are used in risk assessments:

hazard – something with the potential to cause harm
probability – the likelihood that the hazard will cause damage or harm
risk – a compound of the probability and the severity of the resulting damage or harm
danger – the state of being at risk
extent of the risk – a measure of the number of people likely to be affected and the severity of damage or harm, i.e. the consequences.

3.4.2 Objective

The object of a risk assessment is to identify hazards so that action can be taken to eliminate, reduce or control them before accidents occur that cause injury or damage.

3.4.3 Strategy

To achieve that objective and for risk assessments to be effective and workable, they need to be approached systematically. The following steps outline a logical and systematic approach:

1. Define the task or process to be assessed.
2. Identify hazards.
3. Eliminate hazards or reduce to a minimum.
4. Evaluate the residual risks.
5. Develop precautionary strategies.
6. Train operatives in new work methods.
7. Implement precautionary measures.
8. Monitor performance.
9. Review periodically and revise as necessary.

---

* The method of recording is not specified but could be by diary note, special form or on disc. It is prudent to record all items viewed – the safe and the hazardous – for future reference and evidence of the extent of the assessment.

Considering each of these steps separately:

1. Identify hazards.

   Techniques:
   - safety inspections (a general safety survey of the workplace)
   - safety tours (identify hazards along a fixed pre-determined route)
   - safety sampling (check for only one type of hazard, then later repeat for other hazards)
   - safety audit (a count of the numbers of the different hazards found for comparison with similar earlier or later audits)
   - environmental surveys
   - accident reports
   - reports of near misses or 'close shaves'
   - comments from employees.

2. Eliminate or reduce hazards.

   Possible action:
   - eliminate hazardous operation or material (the matter then ends since the hazard no longer exists)
   - for those hazards that cannot be eliminated:
     - develop safer working methods
     - use alternate less hazardous materials.

3. Evaluate the residual risks.

   Considerations:
   - level or extent of hazard faced
   - time of exposure
   - number of people exposed
   - probability of event happening.

   Qualitative evaluation:
   - based on personal estimation
   - essentially subjective
   - can be given numeric values (see below).

   Quantitative evaluation:
   - based on published data of failure rates
   - often quoted as probabilities.

4. Develop precautionary strategies.

   Techniques:
   - elimination of hazardous equipment, materials, substances or method of working
   - substitute safer equipment, materials, substances or working methods
   - prevent exposure or contact by use of bulk supplies or containment
   - control exposure or contact by restricting access to or time of contact with substance
   - provide PPE as a last resort.

5   Train operatives:

   - in new working methods
   - in proper use of precautionary measures.

6   Implement precautionary measures.

7   Monitor performance.

   Ensure the following:

   - precautionary measures/working methods are being used
   - the precautionary measures are effective
   - the new work methods have not created new hazards
   - possible weaknesses in the measures are highlighted.

8   Review and revise.

   - to ensure methods still effective
   - to bring precautions up-to-date
   - whenever work methods or materials change
   - when existing assessment becomes no longer effective.

Figure 3.1 illustrates this process diagrammatically.

3.4.4 Risk rating

Where a number of residual hazards remain there may be a need to determine priorities for dealing with them. This can be done either subjectively based on the assessor's knowledge of the processes or by evolving, for each residual hazard, a numeric value or risk rating based on a number of factors:

Factor	Extent	Value
Hazard	• unlikely to cause injury	1
	• may cause minor injury	2
	• could cause injury requiring first aid	3
	• could cause injury requiring medical treatment	4
	• could cause major injury	5
	• life threatening, possibility of fatality	6
Probability	• most unlikely	1
	• remote possibility	2
	• reasonably possible	3
	• fairly likely	4
	• very likely	5
	• almost certain	6
Severity	• negligible injury	1
	• minor injury	2
	• major injury	3
	• multiple injuries	4
	• single fatality	5
	• multiple fatalities	6

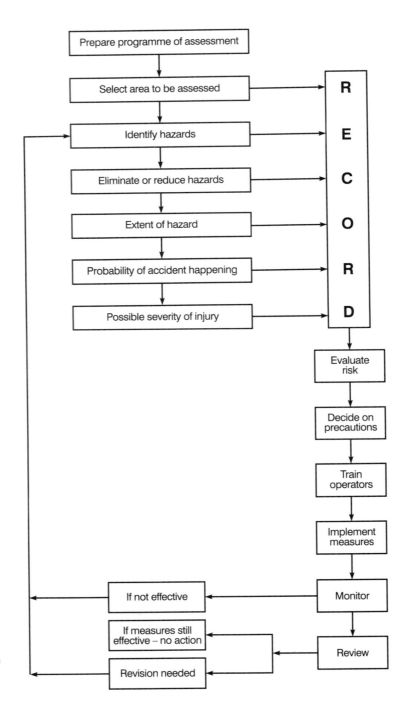

Figure 3.1 Diagram of the process of risk assessment

The risk rating is obtained by multiplying together the values given to each of the factors:

hazard value × probability value × severity value = risk rating

This gives a numeric value to each of the hazards which, as well as highlighting the highest risk, provides the basis for a priority rating.

3.4.5 Design risk assessment

The above deals with risk assessment in respect of existing plant and premises. However, risk assessments at the design stage are becoming increasingly demanded. These require specialist knowledge of and experience in the operation of the particular plant and are often carried out by multi-discipline teams.

Typical techniques include:

- 'WHAT-IF' technique of assessing the likely effects of foreseeable faults.
- Fault Tree Analysis (FTA), developing a logic diagram to trace back possible faults.
- Hazard Analysis (HAZAN), analysing the effects of possible faults.
- Hazard and Operability Studies (HAZOPS), involving a multi-discipline group considering the effects of faults.
- Failure Mode and Effect Analysis (FMEA), studying the effects of component failures.

3.4.6 Legislation

Existing legislation that has a risk-assessment requirement includes:

- *The Control of Asbestos at Work Regulations 1987*
- *The Control of Lead at Work Regulations 1980*
- *The Control of Substances Hazardous to Health Regulations 1994*
- *The Management of Health and Safety at Work Regulations 1992*
- *The Personal Protective Equipment at Work Regulations 1992*
- *The Manual Handling Operations Regulations 1992*
- *The Health and Safety (Display Screen Equipment) Regulations 1992*
- *The Provision and Use of Work Equipment Regulations 1992* in its requirement to determine safety of use 'so far as is reasonably practicable'
- *The Supply of Machinery (Safety) Regulations 1992* in the 'essential safety requirement' to *eliminate or reduce risks as far as possible.*

3.4.7 Sources of information

Major sources of information are the Approved Codes of Practice associated with the above Regulations. These are available from HSE Books:

COP 2	Control of lead at work	£3.90
L 5	General COSHH ACOP and Carcinogens ACOP and Biological Agents ACOP (1996 edition): Control of Substances Hazardous to Health 1994	£7.50
L 21	Management of Health and Safety at Work Regulations 1992. Approved Code of Practice	£5.00

44 Management responsibilities

L 22	Work equipment. Provision and Use of Work Equipment Regulations 1992. Guidance on the Regulations	£5.00
L 23	Manual handling. Manual Handling Operations Regulations 1992. Guidance on the Regulations	£5.00
L 25	Personal Protective Equipment Regulations 1992. Guidance on the Regulations	£5.00
L 26	Display screen equipment at work. Health and Safety (Display Screen Equipment) Regulations 1992. Guidance on the Regulations	£5.00
L 27	The control of asbestos at work. Control of Asbestos at Work Regulations 1987. Approved Code of Practice 2nd edition	£5.00
HS(G) 97	Booklet 'A step by step guide to COSHH assessments'	£5.00
INDG163	5 steps to risk assessment	single copy free

3.5 Techniques of hazard identification

The identification of hazards before they cause an accident is central to all accident prevention activities. However, hazard identification is not an exact science but a subjective activity where the measure of the hazard identified will vary from person to person depending on their experiences, attitude to risks, familiarity with the process, etc. By repeating, or employing a range of, identification techniques the number of residual hazards will be reduced. It is doubtful if they will all ever be totally eliminated.

The findings of each inspection should be recorded so they can be referred to when deciding remedial action needed and for comparison with previous inspections.

There are a number of identification techniques from which to select the one that is likely to be most effective in a particular organization or which will provide the information required in respect of a particular process. They include:

1 Safety surveys

- Sometimes called safety inspections.
- Entail a general inspection of the whole work area.
- Tend to be less detailed than other techniques.
- Do give an overall picture of the state of accident prevention across the particular work area.

2 Safety tours

- Inspection is restricted to a predetermined route.
- Need to plan subsequent routes to ensure complete coverage of work area.
- Reduces the time taken by each inspection.

3 Safety sampling

- Looks at only one aspect of health or safety.
- Concentrates the mind and identifies more detail.
- Need to plan a series of samplings to cover all aspect of health and safety.

Management responsibilities

4 Safety audit

- Detailed inspection of the workplace.
- Seeks to identify all types of hazard.
- Numbers of each type of hazard identified can be recorded.
- Can be developed to give a numerical score to measure the 'safety health' of the organization.
- By repeating the audit a measure of improvement, or otherwise, in safety health can be obtained.
- Can be time-consuming.

5 Environmental checks

- Based on measurements of concentrations of chemicals in the atmosphere.
- Can identify possible health hazards faced in workplace.
- Recording of sequential readings can show improvements or otherwise.
- Checks by 'grab sampling' are not very accurate and can be expensive.
- Electronic instruments expensive to buy but give instantaneous accurate reading.
- Electronic instruments can be used continuously over a long period.

6 Accident reports

- Post accident recording.
- Need to include minor as well as lost time injuries.
- Information obtained from accident report.
- Report should give indication of preventative action needed.

7 Near-miss reports

- Reports of incidents that in slightly different circumstances could have caused an accident.
- Needs the right safety culture to be effective.

8 Feedback from employees

- Can be formally through a safety committee or informally to supervisor.
- Needs a 'no-blame' culture to encourage employees to report matters.
- Employees often know and say what needs to be done.
- Needs feedback on action taken to retain management credibility.

3.5.1 Typical matters for inspections

- Welfare facilities:
  - canteen
  - toilets
  - first aid
  - smoking arrangements

- Fire precautions:
  - extinguishers
  - escape routes
  - alarms and fire drills
  - non-smoking areas

- Machinery:
  - guarding
  - following agreed system of work
  - state of machinery
  - reports of statutory examinations
  - compliance with legislative requirements

- Working conditions:
  - temperature
  - lighting
  - cleanliness and housekeeping
  - fumes and dusts
  - general decoration

- Access and gangways:
  - well marked
  - not encroached on
  - surface condition
  - adequately lit.

3.6 Organization for safety at work

The object of organization within an enterprise is to provide the means by which the aims of that enterprise can be achieved. As far as safety is concerned, it must make provision for, not only its own employees, but also for visitors to the premises, contractors working on the premises, invitees, neighbours and members of the public who may be affected by the way the enterprise operates.

3.6.1 Techniques

Techniques an organization can utilize to ensure its efforts are effective include:

- good communication
- consultation
- commitment from all
- generating identity with the organization
- involvement and participation
- work and job design
- competitive payment systems
- commitment to quality
- satisfying the customer.

3.6.2 Types of organization

Formal – the organization's structure is dictated by directors as being that which is necessary to meet commercial aims.

Informal – casual groups of individuals band together, linked by common interests. They decide amongst themselves how they will behave in the environment in which they work and fix their own work targets, often in

Management responsibilities

conflict with the demands of the formal organization. Compliance with the norms set by these informal groups is often given priority over those imposed by the formal organization.

Official – i.e. governmental departments; in health and safety these are the Health and Safety Commission and Executive.

Professional – the Institution of Occupational Safety and Health, the Chartered Institute of Environmental Health Officers, British Occupational Hygiene Society, etc.

3.6.3 Roles in the organization

Within an organization the health and safety role played varies according to the person's position within the hierarchical structure. Thus:

- Managing director
  - sets the tone for the organization by his attitude, commitment and involvement
  - controls resources and ensures that facilities are available for achieving the safety policy aims
  - makes resources available for health and safety matters and ensures they are used effectively

- Production manager
  - responsible for organizing the work and ensuring it is carried out safely
  - consults with the employees on health and safety matters
  - agrees safety rules and practices
  - ensures those safety rules are followed
  - provides the means (labour, materials and finance) to achieve and maintain a safe workplace
  - chairs the safety committee

- Safety adviser
  - advises managers on all matters of health and safety
  - organizes safety committee meetings but does not take the minutes
  - acts as contact with outside safety organizations such as HSE, RoSPA, BSC, local safety groups

- Foreman/chargeman
  - front-line manager
  - checks that safety rules are being followed
  - ensures machinery and equipment are safe to use
  - initiates discipline for breaches of safety rules
  - initiates requests for safety work

- Safety representative
  - represents employees on safety matters
  - carries out inspections and investigations with the agreement of the manager
  - ensures employees follow the safety rules

- Safety committee
  - considers reports from the manager and safety adviser on safety matters

- comments on safety standards and practices and makes recommendations for improvements
- keeps employees informed on safety matters

- The company safety organization and performance should reflect the organization and performance elsewhere in the company, especially where BS EN ISO 9001 registration exists.

3.6.4 Safety organizations

- The Institution of Occupational Safety and Health (IOSH)
  - the recognized professional body
  - sets professional standards for practising safety advisers
  - represents the interests of the safety practitioner

- The National Examination Board in Occupational Safety and Health (NEBOSH)
  - the recognized examining body for occupational health and safety subjects
  - independent and self-financing

- The Royal Society for the Prevention of Accidents (RoSPA)
  - largest UK safety organization
  - covers safety in all work and leisure activities
  - provides training and safety advice
  - organizes major safety exhibitions

- British Safety Council (BSC)
  - independent safety organization
  - provides training and safety advice
  - has strong safety lobby

- British Standards Institution
  - not strictly a safety organization but many of its standards contain safety requirements

- Industry safety bodies
  - organized within and by particular industries
  - often in co-operation with the HSE through Industry Advisory Committees
  - set standards particular to their industries
  - voluntary and rely on employers to implement the standards.

3.7 Promoting health and safety in the workplace

There are a number of techniques that can be employed to improve and promote effective levels of health and safety in the workplace that complement legislative requirements and are good industrial and commercial practice. They aim to increase awareness of the need for high standards of health and safety at work.

Typical techniques include:

- Evaluation of safety knowledge
  - Complete Health and Safety Evaluation (CHASE)
  - International Safety Rating System (ISRS).

Management responsibilities

- Risk assessment
  - identify hazards and remove them or take appropriate precautions.
- Monitoring safety standards
  - safety inspections and surveys that are general in nature and cover whole workplace
  - safety tours that follow a predetermined route and note safety items
  - safety audits comprising detailed examination and quantification of safety items
  - safety sampling that looks only at one specific aspect of health or safety.
- Communicating the safety message by:
  - poster
  - news sheet
  - tool box talks
  - personal example.
- Using safer processes or materials.
- Including health and safety as inherent part of skill training.
- Keeping all plant well maintained
  - planned maintenance.
- Developing and using safe systems of work.
- Ensuring supervisors are trained and competent in health and safety matters.
- Practice of emergency drills and procedures.
- Providing good working conditions and environment.

3.7.1 Safe systems of work

A **safe system of work** is a considered method of working that takes proper account of the potential hazards to employees, and others such as visitors and contractors, and provides a formal framework to ensure that all steps necessary for safe working have been anticipated and implemented.

All systems of work should be safe, but where the risk is such that unambiguous working instructions are necessary, the system of work should be in writing.

Typical procedure for developing a safe system of work:

1. Identify hazards from:
   - energy
     - electricity
     - steam or high pressure hot water
     - compressed air
     - hydraulic systems
     - compressed springs (in cylinders, robots, etc.)
   - materials
     - corrosives
     - asphyxiant gases
     - flammables and explosives
     - toxic substances
   - plant
     - machinery
     - cranes and lifting equipment
     - internal transport
   - dangerous places
     - working at heights
     - in confined spaces
     - strange environment.

2 Remove danger
- change process and/or materials.

3 Provide protection
- guards
- personal protective equipment.

4 Develop safe system of work
- in writing
- use:
  * locking off
  * permit-to-work.

5 Provide suitable training.

6 Provide special equipment
- harnesses
- breathing apparatus/masks
- ear muffs/plugs
- safe working platforms.

7 Monitor that the system is being followed.

A useful mnemonic device is 'IRPSTEM':

I – Identify hazards
R – Remove dangers
P – Provide protection
S – Safe system of work
T – Training
E – Equipment
M – Monitor.

3.7.2 Permit-to-work system

Certain operations present higher than normal risks and require stricter controls. In these cases a 'permit-to-work' system should be implemented. A **permit-to-work** procedure is a formal written system of authorization used to control certain types of work which are potentially hazardous.

The permit-to-work must be in writing and each part authorized by a suitably responsible person. The essential parts of a permit-to-work are:

1 Statement (description) of the work to be undertaken and the safety precautions to be implemented before work can commence.
2 Confirmation that the precautions have been taken and it is safe to commence the work.
3 Statement that the work is complete and it is safe to return the plant to production.

In carrying out this procedure it is essential that:

- there is clear understanding of who:
  - may authorize each part
  - is responsible for specifying the necessary precautions.
- proper training and adequate instructions are given in:
  - the work to be carried out
  - the procedure to be followed in the issue and use of the permits.

X Y Z Company Limited
PERMIT-TO-WORK

NOTES:
1. Parts 1, 2 and 3 of this Permit to be completed before any work covered by this permit commences and the other parts are to be completed in sequence as the work progresses.
2. Each part must be signed by an Authorized Person who accepts responsibility for ensuring that the work can be carried out safely.
3. None of the work covered by this Permit may be undertaken until written authority that it is safe to do so has been issued.
4. The plant/equipment covered by this Permit may not be returned to production until the Cancellation section (part 5) has been signed authorizing its release.

PART 1 DESCRIPTION

(a) Equipment or plant involved _____

(b) Location _____
(c) Details of work required _____

Signed _____   Date _____
person requesting work

PART 2 SAFETY MEASURES

I hereby declare that the following steps have been taken to render the above equipment/plant safe to work on: _____

Further, I recommend that as the work is carried out the following precautions are taken: _____

Signed _____   Date _____
being an authorized person

PART 3 RECEIPT

I hereby declare that I accept responsibility for carrying out the work on the equipment/plant described in this Permit-to-Work and will ensure that the operatives under my charge carry out only the work detailed.
Signed _____   Time _____   Date _____

Note: After signing it, this Permit-to-Work must be retained by the person in charge of the work until the work is either completed or suspended and the Clearance section (Part 4) signed.

PART 4 CLEARANCE

I hereby declare that the work for which this Permit was issued is now completed/suspended* and that all those under my charge have been withdrawn and warned that it is no longer safe to work on the equipment/plant and that all tools, gear, earthing connections are clear.
Signed _____   Time _____   Date _____
* delete word not applicable

PART 5 CANCELLATION

This Permit-to-Work is hereby cancelled
Signed _____   Time _____   Date _____
being a person authorized to cancel a Permit-to Work

Figure 3.2 A Permit-to-Work

- the work is monitored to ensure the laid down procedures and methods are being followed.

The objects of the procedure are to:

- reduce to a minimum the risk of injury or ill-health to those carrying out the work
- ensure proper authorization for the procedures and work
- make clear to those carrying out the designated work:
  - the exact identity, nature and extent of the job
  - the hazards faced, the precautions to be taken
  - any limitations on the extent of the work or time allowed
- ensure that the person in charge of the plant or area is aware of all the work that is to be done
- provide a record of the nature of the work, the precautions taken and the people involved
- provide a formal hand-back procedure to ensure that the part of the plant affected by the work is in a safe condition to return to production
- A typical Permit-to-Work is shown in Figure 3.2.

4   Human resources

One part of the work environment that has, in the past, been taken very much for granted is the workforce, even though they are probably the most costly component in a company's budget. In recent years this attitude has changed and there is now a great deal of attention focused on and legislation aimed at protecting employees, not only from the dangers met at work but also in the security of their employment and in giving them a greater say in the running of the enterprise.

The following sections deal with those aspects of health and safety that are related specifically to employees in their various roles.

4.1 Health and safety training

Obligations are placed on employers by s. 2(2)(c) of HSW:

... to provide such information, instruction, training and supervision as is necessary to ensure, so far as is reasonably practicable, the health and safety at work of employees.

This is enlarged on in Regulation 11 of the *Management of Health and Safety at Work Regulations 1992* which requires employers to:

- take account of the ability of employees to perform their tasks
- provide training when:
  - employees:
    * first join the company
    * are transferred to a different job
    * are given changed responsibilities
  - working methods of existing equipment are changed
  - new equipment is introduced
  - new technology is introduced
  - new materials are used
  - the system of work is changed
- give revision training periodically
- give training during working hours.

Where contract staff are hired on a temporary basis and are put to tasks for which training is considered necessary for full time employees, the employer is required to give those temporary staff the same training.

Where contractors are brought on to the premises to carry out work, they too must be trained in any particular risks they may face and in the techniques to avoid those risks.

54 Human resources

4.1.1 Training content

Induction training

- For all employees:
  - to cover all aspects of employment including:
    * fire precautions and evacuation drill
    * details of company products
    * tour of the premises to identify location of facilities
    * occupational health facilities such as first aid arrangements
    * accident prevention activities
    * safety rules to be obeyed
    * fire and security arrangements
    * who to contact if in doubt
- For contractor and his employees:
  - local safety rules to be obeyed
  - any special hazards in their area of work
  - safe systems of work and permit-to-work systems procedures
  - emergency and evacuation procedures
  - allowed access routes
  - permitted use of welfare facilities – canteen, toilets, first aid, etc.

On-going health and safety training

- For all employees:
  - re-affirmation of safety rules
  - hazards likely to be met in the work and techniques for avoiding them
  - safety devices and how to use them
  - issue, use and maintenance of PPE
  - re-affirmation of emergency and evacuation procedures
  - action in the event of an accident
  - procedure on identifying a hazard
- For supervisors
  - more detailed instruction on legal requirements – HSW, regulations and approved codes of practice
  - common law duties of care
  - techniques of risk assessment
  - getting benefit from safety inspections
  - accident investigation
  - fire precautions and special responsibilities
  - need for personal and special hygiene requirements
  - safety committee and the role of the safety representative
  - industrial relations in health and safety
- For managers
  - interpretation of health and safety legislation
  - economics of safety:
    * cost of accidents
    * employer's liability and accident claims
    * risk management
    * cost benefits of safety and provision of safeguards
  - safety culture
  - safety inspections and audits
  - monitoring safety performance
  - motivating for safety
  - integration of safety as inherent part of all work
- For safety representatives (whether union-appointed or voluntary)
  - their role in health and safety in the workplace

- their rights regarding:
  * special training
  * time to carry out inspections
  * investigation of accidents
  * membership of safety committee.

4.1.2 The training process

Before training is given, the need for it should be assessed and the programme of training tailored to meet the identified needs by:

- analysing the training needs
- drawing up a training plan
- developing a training programme
- setting training objectives (what the programme should achieve)
- briefing speakers/trainers
- preparing training aids and visual aids
- implementing the programme as a training course
- evaluating the effectiveness of the training (de-brief)
- revising the programme, speakers, training aids for the next course.

4.1.3 Training techniques

The particular technique used should match the level of course being given but could include:

- lectures and talks
- videos and films
- role-playing by course members
- case studies – report back to course
- syndicate discussions
- practical exercises either out on site or using a table-top model
- on-the-job.

Training is an investment for the future whether it be to improve skills, utilize the latest technology, give employees greater job satisfaction or to ensure they go home in one piece.

4.2 Young persons at work

Children:

- under 13 years of age are prohibited from employment
- between the ages of 13 and 16 may not be employed in industrial processes unless on a training scheme approved by the local authority
- aged between minimum school leaving age (around 16th birthday) and 18 years, generally known as 'young persons', may be employed but under strictly controlled conditions.

Youngsters leaving school to start work are vulnerable because:

- they are entering a strange environment that is alien to what they have known so far
- they have no built-in recognition of the dangers of machinery or equipment

56 Human resources

It's never too early to start safety training

- they are at a stage of physical development that leaves them susceptible to a range of chemical substances and physical hazards that may cause lifelong disabilities or effects.

The Health and Safety (Young Persons) Regulations 1997 require employers employing young persons to:

- Assess risks they may face *before* they start work.
- Provide information to parents about possible risks faced and precautionary measures.
- Make allowance for inexperience, immaturity and lack of awareness of hazards.
- Prevent young persons from using high risk machinery or processes except:
  - where necessary for their training and
  - risks reduced as far as is reasonably practicable and
  - adequate and competent supervision is provided.

The above restrictions do not apply to part-time work in a concern owned by and employing only members of the same family.

Youngsters taken on or given employment under any of the Goverment's 'training for employment' schemes should be treated as employees and given the same training and supervision as full time youngsters.

Human resources **57**

4.2.1 Problems faced by young persons starting work

In making the transition from school to work young persons need to adapt to:

- change from $5\frac{1}{2}$-hour day to $7\frac{1}{2}$- or 8-hour day
- monotony and possibly boredom until work understood
- lack of breaks every 40 minutes
- unfamiliar equipment and machinery
- unfamiliar hazards
- safety rules to follow
- safety devices to learn about
- protective clothing.

4.2.2 Training for young persons

- Same induction training, as for all employees, plus emphasis on:
  - need for good housekeeping
  - no running
  - no horse play or practical jokes
  - no short cuts:
    * in the job
    * through the workplace
  - role of supervisor as trainer
  - if in doubt, ask
  - where to get information and advice
  - explain 'why fors'
  - follow the example of the supervisor.

4.2.3 Supervision

Both statute and common law place a special duty on employers to ensure young workers are properly and adequately trained and supervised. Supervision is crucial to ensuring the safety of youngsters, especially in their early days at work and when being put to work on machinery of any sort. The supervisor should remain in the vicinity of the youngster to keep an eye on what is being done and to be instantly available to give advice or to prevent bad habits from developing.

4.2.4 Special risks

Where work involves any of the following processes or materials, special attention should be paid to training and protecting young persons:

- carriage of dangerous goods
- dangerous (heavy) metals
- dangerous machinery
- driving work vehicles, cranes, etc.
- explosives
- high fire-risk processes
- ionizing radiations
- lead
- manual handling
- power presses

- power-driven machinery
- toxic substances
- woodworking machines (effectively prohibited unless under supervised training).

In addition, legislation places severe restriction on young persons working in:

- agriculture
- potteries
- asbestos
- nuclear
- certain parts of the chemical industry.

Typical of the machines at which young persons should not work without proper training and supervision were listed in the *Dangerous Machines (Training of Young Persons) Order 1954* as:

1. Machines driven by power:

    - brick and tile presses
    - machines used for opening or tesing in upholstery or bedding works
    - carding machines in use in the wool textile trades
    - corner staying machines
    - dough brakes
    - dough mixers
    - worm pressure extruding machines
    - gill boxes in use in the wool textile trades
    - the following machines in use in launderies:
        - hydro-extractors
        - calenders
        - washing machines
        - garment presses
    - meat mincing machines
    - milling machines in use in the metal trades
    - pie and tart making machines
    - power presses including hydraulic and pneumatic presses
    - loose knife punching machines
    - wire-stitching machines
    - semi-automatic wood-turning lathes.

2. Machines whether driven by power or not:

    - guillotine machines
    - platen printing machines.

While this Order has been revoked, it does provide a useful list of machines at which special precautions are necessary if young persons are employed.

Particular care and attention needs to be paid in the training of youngsters starting work to correcting any bad working habits since they will be carried throughout their working life.

4.3 Joint consultation

In endeavours to ensure high levels of health and safety in the workplace, use must be made of all the know-how available. Where matters relating to the actual work areas are concerned, the people best qualified to comment are those who work there. In many cases they see hazards and problems and get round them long before they reach a manager's notice. Also where a problem arises concerning particular aspects of the workplace, the local workers are well placed and well qualified to offer helpful suggestions for its solution. This potential needs to be tapped.

In addition, if employees are consulted before changes are made they will accept them that much more readily.

Joint consultation can take many forms and routes depending on the culture and environment in the workplace. Rights to appoint safety representatives have been given to recognized trade unions who are under no pressure to exercise this right which remains should they feel the need to call on it later. These rights are embodied in the *Safety Representatives and Safety Committees Regulations 1977*.

Similar rights exist for volunteer or management appointed safety representatives under the *Health and Safety (Consultation with Employees) Regulations 1996*. It has been found that with good industrial relations and an effective voluntary safety committee, the unions are happy not to exercise their rights of appointment.

Joint consultation

4.3.1 Safety representatives

The rights of safety representatives include:

r.3 The right to be consulted on:

- the introduction of measures effecting health and safety
- arrangements for appointing safety adviser
- arrangements for appointing fire and emergency wardens
- the provision of health and safety information relevant to those represented and required by various laws
- the provision of health and safety training
- health and safety implications of the introduction of new technologies.

r.5 The right to be provided with sufficient information:

- to enable them to carry out their functions
- on accidents that have happened, but not where:
  - an individual may be identified
  - it could prejudice the company's trading
  - the matters are the subject of litigation
  - it is against national security
  - it would contravene a prohibition imposed by law.

r.7 The right to be allowed time off to:

- receive training
- carry out a representative's functions.

r.6 The functions of a safety representative are to:

- make representations to the employer on hazards and incidents that affect employees health and safety
- raise with the employer other matters affecting the health and safety of employees
- be a contact with and receive information from HSE inspectors
- investigate potential hazards and accidents.

In addition, union-appointed safety representatives can:

- investigate complaints by employees concerning health and safety matters
- attend meetings of the safety committee
- carry out:
  - subject to giving suitable notice to, and getting the agreement of the employer, inspections of the workplace:
    * at suitable intervals
    * when there have been substantial changes in workplace
    * following notifiable accident, dangerous occurrence or disease.

4.3.2 Safety committee

One of the most important vehicles for joint consultation is the safety committee.

To be effective safety committees need to:

- be properly constituted
- include representation from the shop floor, supervision and management
- have written terms of reference
- have as chairman someone who:
  - can run a meeting
  - is committed to high standards of health and safety
  - has authority to accept and initiate action on committee's recommendations
- work to an agenda which should be distributed at least a week before the meeting
- have an agreed procedure for raising matters, i.e. not until supervision has been given adequate time to take any necessary action and nothing has resulted
- develop a system for measuring its effectiveness by:
  - recording items raised
  - list new items raised at each meeting
  - noting number of jobs completed since last meeting
  - getting explanations for delays in completion
  - agree programmes of inspections and require a report back.

Note that it is not the function of the safety committee to bypass line management. Any safety hazard or dangerous occurrence identified must be reported to supervisors for them to initiate corrective action. This is the responsibility for which they have been given authority. No matter should be taken to the safety committee before the supervisor concerned has been given an opportunity to correct it. Only if no action results may the matter be referred to the committee.

The agenda can include an item on company affairs where the chairman can keep the committee informed of company developments and proposed actions.

Safety representatives and safety committees have a positive role to play in stimulating high levels of awareness and generating high standards of health and safety in the workplace.

4.4 Industrial relations in health and safety

The term 'industrial relations' includes any matters that influence the relations between employer and employee particularly as they affect conditions of employment. In that respect, health and safety are major features of employment conditions.

Safety performance can be a good indicator of the state of industrial relations. Conversely, good industrial relations generate high safety performance.

Foundations for good industrial relations include:

- good communications – both ways
- honesty in dealing with people
- openness of approach
- clear, well understood and agreed safety rules
- well defined and understood grievance and disciplinary procedures
- discipline which, when used, must be seen to be fair

- good consultation arrangements that are seen to work
- trained and competent supervision
- visible action following complaint/request or an explanation of why not.

With the proliferation of health and safety laws and standards, questions of discipline may arise that revolved round health and safety issues. It is necessary to be aware of the recognized procedures and practices in this field to ensure speedy and effective settlement of matters. These procedures and practices stem from the vast body of industrial relations law which exists today. That law includes, *inter alia*:

Employment Protection Act 1975 – established the Advisory Conciliation and Arbitration Service (ACAS) which produced a Code of Practice 1 on 'Disciplinary practice and procedures in employment' recommending that disciplinary procedures should:

- be in writing
- specify to whom they apply
- require that matters are dealt with quickly
- outline the disciplinary action that can be taken
- specify the levels of management that have the authority to take the various forms of disciplinary action, ensuring that immediate superiors do not have power to dismiss without reference to more senior managers
- ensure individuals are informed of complaints against them and that they are given an opportunity to state their case before decisions are reached
- give individuals the right to be accompanied by a union representative or by a colleague
- except for gross misconduct, ensure that no employee is dismissed for a first breach
- ensure cases are fully investigated before disciplinary action is taken
- ensure individuals are given an explanation for any penalty imposed
- specify appeal procedure.

The Disabled Persons (Employment) Act 1944 – requires the employment of a certain percentage of registered disabled persons. Care must be taken that, because of their disability, their health and safety is not put at risk, especially in an emergency.

The Equal Pay Act 1970 – requires equal treatment for men and women in the same employment and doing 'like work'.

Sex Discrimination Acts 1975 and *1986* require that a person is not treated differently because of his or her sex or marital status. It applies equally to men and women. It removed restrictions on overtime and night work for women but allowed discrimination on health grounds.

Race Relations Act 1976 is similar to the Sex Discrimination Acts but is aimed at ensuring that no one will be prejudiced in employment by virtue of their race, colour, nationality, ethnic or national origins. Issues arise with Sikhs and the wearing of hard hats and employment in the food industry.

Similarly, *The Disability Discrimination Act 1995* requires that no one is discriminated against because of their disability.

The Trade Union Reform and Employment Rights Act 1993 made extensive changes to industrial relations laws including:

> - giving increased maternity rights for women regardless of the length of service
> - providing employment protection for safety representatives, whether union-appointed, employer-appointed or voluntary, where health and safety was an issue.

In practice, should discipline become necessary, the steps outlined below form a sound procedure to follow:

1. Give an oral warning with an opportunity to improve/correct the fault complained of, offer additional training if relevant. Also, allow employee a chance to explain his side of the complaint.
2. If no improvement, give a first written warning including a statement of the possible consequences of not improving, i.e. dismissal.
3. Give a second written warning, repeating the statement given in first letter.
4. Give a final written warning including a statement about possible dismissal.
5. If no improvement, dismiss.

The state of industrial relations reflects the culture of an organization and the attitudes, involvement and commitment of senior managers. Employees tend to react to the behaviour of those managers so an important facet of good industrial relations is the setting of a good personal example by the senior people in the organization.

4.5 Human factors in health and safety

In any activity involving human beings the effectiveness with which it is carried out depends very much on the way in which the individuals concerned look at what is to be done. Approaches can range from supreme enthusiasm that brooks no setbacks to downright indifference that looks for any reason not to do anything. The difference between these two extremes is in the attitude and manner in which the individual approaches the activity. This may be due to internal factors that are very personal to the individual or to external factors such as the circumstances or the environment in which the individual finds himself and over which he has no control. In the occupational field these factors and any others that affect the interface between employees and their work are referred to as *human factors*.

It is being increasingly recognized that many of the accidents that occur at work are a direct result of human factors, i.e. the accident has a behavioural cause rather than the failure of a mechanical part or a shortcoming in a system of work.

Occupational **human factors** refer to any matters that influence a person's approach to work and the ability to carry out the job tasks. Those influences can occur at any point in the person's daily activities whether in the home, at work, at social gathering or during leisure activities. Human factors are one of the many facets of behavioural science.

This section, for simplicity and convenience, will consider the influences on human factors under four headings:

1. What human factors cover.
2. Positive factors – some factors that can improve attitudes to work.
3. Negative factors – some factors likely to increase risks at work.
4. Personal factors.

1. Human factors cover:

   - the attitude of employees to their work
   - relationships between employees and their work groups
   - interaction between an individual and the job or work environment
   - individual capabilities and fallibilities (human error)
   - personal behaviour of individual
   - the extent of training and instruction provided
   - the design and condition of plant and equipment
   - rules and systems of work – whether reasonable and acceptable.

2. Positive factors:

   - a managerial environment that sets a proper safety culture
   - matching the individual to the job or machine
   - on-going training covering:
     - skills to carry out the work
     - knowledge of processes
     - use of work equipment
     - company plans and aspirations
   - providing equipment that is:
     - safe
     - kept in good condition
     - can be adjusted to operator's capabilities of speed, size, dexterity, etc.
     - ergonomically designed
   - having performance goals that are:
     - realistic
     - attainable
     - understood
     - acceptable
   - discipline which is seen to work and be fair
   - provision of adequate information about:
     - the job to be done
     - the company
     - work targets
   - monitoring performance and communicating results
   - having a system of 'feedback' to enable employee's ideas to be utilized and recognized
   - ensuring agreed rules and procedures are followed.

3. Negative factors

   - lack of training in tasks
   - 'macho' attitude towards rules and safeguards
   - ignoring or bypassing safeguards and taking short-cuts to increase take-home pay

- ignorance of what is going on
- ignorance or misunderstanding of what needs to be done
- failure to communicate or instruct properly
- bad design and layout of plant and equipment that does not take account of human limitations, both physical and mental (ergonomics)
- lack of clear direction.

4 Personal factors:

- individual attitudes to the job and work
- degree of personal motivation from the work
- whether the training received satisfies personal needs
- perception of the value of role in the organization
- ability to match demands of the job
- seeing the work as a challenge.

In their publication no: HS(G)48, 'Human factors in industrial safety', the HSE look at a range of factors and review a number of circumstances under the headings:

- the organization
- the job
- personal factors

with the aim of stimulating interest rather than answering questions.

Building on the positive factors and working to eliminate the negative factors will make a major contribution towards a more stimulating and safer working environment.

4.6 Insurance in health and safety

No matter how prosperous a company or how substantial its position, unexpected events can occur that can be very costly to resolve and put enormous financial strain on its resources. There are very few companies large enough to be able to absorb, within their own resources, the sort of cost involved in litigation and damages, and the almost universal approach is to spread the risk to institutions set up to handle them, the insurance companies.

Not only is it the employer who looks for recompense following an incident, but so does the employee. While the employee may be able to make a claim for damages against his employer, this is not always so and other avenues for obtaining compensation exist for him through the social security legislation.

This section deals with the sources of recompense for an employee as well as the role of insurances companies in providing sources of finance to the employer to meet unexpected demands resulting from accidents. It also looks at the influence the insurance companies have in stimulating high standards of health and safety.

4.6.1 Sources of recompense for the employee

1 From the State

   (i) *Workmen's Compensation Act 1925*:
        - brought in to ensure that those injured at work received some compensation, but there were restrictions to its application

(ii) *National Insurance (Industrial Injuries) Act 1946*:
- has largely overtaken the Workmen's Compensation Act
- covers a broad range of benefits
- provides payments for absences due to other than industrial causes

(iii) *Social Security Act 1989*:
- arranged for the employer to pay State benefits and to reclaim them

(iv) *Social Security (Industrial Injuries) (Prescribed Diseases) Regulations 1980*, as variously amended:
- lists those conditions that attract benefit
- requires exposure for a period of over ten years
- limits period within which a claim must be made to five years
- specifies industrial processes causing ill-health that can attract benefit.

2 From the employer

A claim for compensation from an employer could be on the grounds of:

- *negligence* if the harm arose because the employer failed to meet a common law duty of care, or
- *breach of statutory duty* if the harm arose because the employer failed to comply with a statutory requirement, or
- both of the above.

If a claim is successful, compensation will only be received if adequate funds are available. This is ensured by employers having Employer's Liability insurance which is a mandatory requirement of:
Employer's Liability (Compulsory Insurance) Act 1969

- Requires employers to carry insurance to cover the likely cost of claims for bodily injury or disease sustained at work (EL insurance).
- Injury or disease must 'arise out of or in the course of employment'.

The 'duties of care' an employer owes to his employees has developed from earlier claims litigation and includes:

- providing a safe place of work
- providing sufficient, safe and suitable plant
- keeping plant well maintained
- providing safe systems of work
- providing adequate and competent supervision
- employing responsible people.

An injury resulting from a breach of any one of these duties of care could be grounds for a claim. (Claims for compensation are civil actions under common law.)

The employer has available to him a number of defences:

- *Volenti non fit injuria* – meaning the injured person consented to run the risk. This defence is unlikely to succeed in the present judicial atmosphere.
- *inevitable accident* – meaning that it occurred in spite of reasonable care by the employer.
- *Contributory negligence* – a partial defence attempting to put some of the blame on the claimant.
- *Res ipsa loquitur* – a plea used if the employer can prove that the accident could have occurred even if he had not been negligent.
- The action was brought outside the permitted time limit, normally three years, but in some cases (diseases) will be allowed within three years of 'date of knowledge' of the cause of the condition.

The EL insurance cover is normally unlimited and extends to include the legal costs as well as any compensation that has to be paid. However, if there has been a prosecution it does not cover the cost of any fine imposed.

EL policy premiums are based on declared payroll figures and cover all those included. Some necessary but voluntary work activities, such as fire fighting training, first-aid practices, etc., are undertaken outside working hours so are not included in the payroll and any injuries occurring are not covered. However, if the individuals concerned are paid a nominal honorarium, and this is included in the payroll return, they will have EL cover for those activities.

Following an accident that is likely to give rise to a claim, the employer notifies the insurers and submits a claim form.

On receipt of a writ claiming damages:

- The writ is *not* acknowledged but sent to EL insurers.
- Insurers take over handling the claim.
- Insurers either handle the case themselves or appoint solicitors to do so.
- Copies of all information about the incident should be sent to insurers including:
  - accident reports
  - form F2508
  - photographs
  - plans of site, plant, accident site, etc.
  - medical certificates.
- Insurer's engineer or solicitor investigates.
- If claimant wants representative to visit site, refer to insurers for decision.
- Insurers negotiate with claimant's solicitor to try for out-of-court settlement since it:
  - saves costs
  - does not set precedents for future compensation levels.
- Insurer decides whether to settle claim or go to court (usually on financial grounds rather than any question of right or wrong).

4.6.2 Other insurance cover

In addition to the mandatory EL cover, it is prudent for an employer to carry insurance cover for a number of other potential risks:

1. Public liabilty:

    - Covers claims for damages from non-employees, e.g. visitors, contractors, neighbours, members of the public.

- Covers the employer's vicarious liability for actions of his employees.

2 Fire insurance:

- Covers losses arising from a fire.
- Includes products, plant, equipment and other assets.
- May be extended to cover loss of profits as a result of a fire.
- Insurers give advice on precautions to reduce fire risks.
- Insurers may place fire precaution conditions to providing cover.

3 Plant and machinery:

- Covers the equipment itself but not injuries arising from its use.
- Insurers have specialist engineers to offer advice on good operating as well as safety techniques and practices.
- This advice may become a condition of cover.
- Insurer's engineers carry out statutory inspections within the policy.

4 Product liability:

- Provides cover against claims for injury, illness or other loss attributed to faulty product.

While the insurance industry's involvement may appear to be low key, it is able to exert an enormous influence to get compliance with required standards through:

- advice on safeguarding of machinery and plant
- advice on techniques for achieving high standards in other operations
- advice on risk management techniques
- reduction of premiums if advice is taken
- the threat to increase premiums if standards are not met, and ultimately to withdraw cover if advice is not followed.

5 Workplace safety

This part is concerned with general safety about the workplace and not with safety in relation to particular processes or machines. It covers access to and egress from as well as movement within the workplace and extends to include welfare facilities.

5.1 Workplace regulations

Requirements for safety in and about the workplace are contained in the *Workplace (Health, Safety and Welfare) Regulations 1992* and its associated Approved Code of Practice no: L 24, which should be referred to for more detailed guidance.

r.3 The Regulations apply to all workplaces except:

- on a ship
- on building sites
- on mineral extraction sites
- in transport vehicle
- on agricultural and forestry land.

r.4 The Regulations place the onus for complying on the 'person in control', i.e. the local manager.

r.5 The workplace, its fixtures and fittings and built-in or attached equipment, devices and systems must be:

- well maintained
- kept clean
- in an efficient state
- in efficient working order
- in good repair preferably backed by a system of planned maintenance with appropriate records.

Maintenance includes:

- inspection
- adjustment
- lubrication
- cleaning
- all equipment, buildings, lighting, escalators, etc.

r.6 Workplace atmosphere:

- to be kept wholesome by
  - opening windows
  - wall- or roof-mounted fans
  - air-conditioning units providing either fresh air or recycled air

70 Workplace safety

Working in confined spaces

- in close or humid areas workers are allowed breaks in a well ventilated area
- where ventilation is necessary to protect employees, the system should:
  - be fitted with a failure alarm
  - supply more than 5–8 litres/second/worker of fresh air
  - be regularly maintained, cleaned and its performance checked
  - not cause draughts
- does not apply to:
  - work in confined spaces, i.e. vats, tanks, sewers, etc., where requirements are laid down in the Confined Spaces Regulations 1997
  - processes where extract ventilation is already required by law, i.e. working with lead, asbestos, radioactivity, dusts and under the COSHH Regulations.

r.7 Temperature during working hours:

- for normal employments: 16 °C (60.8 °F)
- for strenuous work: 13 °C (55.4 °F)

Note: these are minimum temperatures. Actual temperature could be higher to ensure reasonable comfort in the work area, e.g. offices may require a temperature of 20 °C (68 °F) or more.

- in high-temperature work areas such as boiler houses, furnace floors, etc.:
  - provide special cooling arrangements
- in low-temperature work areas such as refrigerators, meat stores, etc.:
  - provide warm clothing
  - a warming room facility
- in the open air or in buildings open to the atmosphere:
  - provide warm clothing in winter
  - the facility of a warming room

- no maximum working temperature is quoted only that it must give reasonable comfort
- high temperatures from sunlight may require the provision of blinds or window shading
- heaters or coolers must not release injurious or offensive fumes into the workplace
- sufficient thermometers to be located in work areas.

r.8 lighting:

- must be 'suitable and sufficient' (no levels are quoted but figures are given in the Chartered Institution of Building Services Engineers' (CIBSE) 'Code of interior lighting')
- wherever possible by natural light
- emergency lighting must be provided where failure of the lighting can give rise to danger
- particular attention should be given to lighting:
  - in rooms with display screen terminals
  - for outside roadways and footpaths
  - on building/construction sites (against glare)
  - in areas with strong shadows
- advice and information on lighting levels from CIBSE.

r.9 Housekeeping:

- the workplace, furniture and fittings must be kept clean
- walls, floors and ceilings must be kept clean
- check for build-up of dust on flat surfaces especially on building structure, roof girders, etc.
- painted walls to be washed and repainted at suitable intervals (say twelve months and seven years respectively) or after changes to the plant or building
- floors should be kept clean by sweeping or washing regularly (at least once a week)
- rubbish must not be allowed to accumulate – it can be a health risk and is a fire risk
- rubbish and waste should be put in suitable containers:
  - oil- or solvents-contaminated waste should be put into fireproof containers such as metal dustbins
- spillages should be mopped up using a suitable absorbent material.

r.10 Space:

- workplaces used for the first time or modified after 1st January 1993:
  - space per employee should not be less than 11 cu. m. (388 cu. ft)
  - space above 3 m. (9 ft 10 ins) should not be considered
- for workplaces in use before 1st January 1993 and not modified:
  - space per employee should be not less than 400 cu. ft
  - space above 14 ft should not be considered

- space taken up by furniture, cupboards, cabinets, etc. may need to be deducted from the space available
- in workplaces with very low ceilings, warning notices should be displayed, and any low beams clearly marked.

r.11 Workstations:

- must be suitable for anyone likely to have to work there
- protected from the weather if reasonably possible
- have well marked emergency exit
- emergency exit routes to be kept clear
- floor kept clean and slip free
- tripping hazards should be removed
- if work platform above floor level it should have safety rail
- if work platform above 2 m. (6 ft 6 ins) a hand rail with intermediate rail and toe board is required
- work in cramped or awkward positions should be for short spells only
- work pieces and materials should be easily reached from work position.

Seating:

- wherever the work allows seating should be provided
- seats should be appropriate for the type of work with back support and foot rests
- typical types of seating:
  - in machine shops, stools
  - for production lines, stools or chairs as appropriate
  - for display screen work, chairs with adjustable height – a foot rest may also be needed
- should be kept in good condition and any damage repaired or seat replaced.

r.12 Floors should:

- be suitable for their purpose, i.e. pedestrian way, traffic route, supporting plant and materials, etc.
- not be over-loaded
- be level and smooth
- not contain potholes, irregularities or depressions likely to cause tripping hazard
- be clear of obstructions with items stowed in designated areas
- not be slippery
- where water is likely to be present have adequate means of drainage
- have separation between traffic and pedestrian ways by either hand railing, barriers or by floor markings
- have a substantial barrier around any openings or holes.

Workplace safety **73**

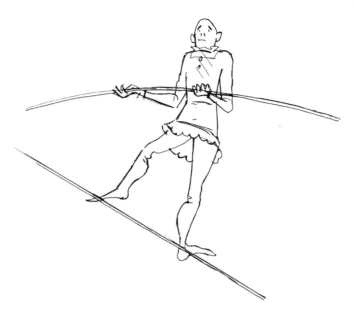

. . . provide means to prevent falls from heights . . .

r.13 Falls from heights:

- to be prevented if likely to cause personal injury
- precautions provided will depend on the particular work:
  - on scaffolding and platforms more than 2 m. (6 ft 6 ins) high a guard rail 1.1 m. (3 ft 6 ins) high with intermediate rail and toe board
  - guard rails are not required for a platform of builder's trestles
  - an operating platform 2 ft high for a machine should have safety rails along its open sides
  - stairs should have a hand rail on both sides whether free-standing or between walls
  - fixed ladders should be fitted with safety hoops
  - for work on roofs:
    * if permanent or frequent, provide fixed walkways with hand rails
    * if temporary, use crawler boards
    * provide edge protection
  - when sheeting lorries either work from a special purpose gantry or if access to the top of the load is necessary, wear a safety harness attached to overhead girder or roof truss
  - if access is necessary to the edge of a platform, the safety rail may be removed but only for the minimum possible time
  - where access is necessary to an elevated platform for the transfer of goods, proprietary safety gates should be used.

Falling objects:

- precautions must be taken to prevent injury from falling objects:
  - by providing toe boards on scaffolding and working platforms
  - bricks and building materials on scaffold platforms should be kept in wire-mesh skips and only removed when being used
  - the provision of rubbish chutes on building sites
  - preventing access to areas below where work is going on
  - racking should be robust and stable with heavy goods stored at low level
  - goods in free-standing stacks should be 'bonded' or the height of the stack restricted (standards developed by industries for stacking their product should be followed, i.e. palletization, end-stacking of paper reels, etc.).

Access ways over vats, etc.:

- walkway must be fitted with guard rails and toe boards on both sides
- vessels containing hazardous substances should have a substantial cover
- vessel includes:
  - tanks
  - pits
  - sumps
  - silos for grain and granular materials
  - hoppers for coal, chemicals, etc.
  - kiers for boiling textile materials
  - hydropulpers.

Guard rails:

- new guard rails
  - top rail 1100 mm (3 ft 6 ins) from walkway
  - toe board 150 mm (6 ins)
  - intermediate rail half-way between top rail and toe board
- old guard rails
  - top rail 915 mm (3 ft)
  - toe board 150 mm (6 ins)
  - intermediate rail (if fitted) half-way between top rail and toe board.

r.14 Glazing:

- transparent or translucent materials used in doors, windows, walls, partitions, etc. at widths greater than 250 mm (10 ins) should be:
  - safety material:
    * polycarbonate
    * glass blocks

* safety glass, i.e. glass that when broken does not splinter
* annealed glass of appropriate thickness
* Georgian wired glass
- large plain sheets of glazing that form a door of an accessway should be marked to make them apparent.

r.15 Skylights and ventilators:

- must be capable of being opened safely
- when open not project and cause a danger to passers-by.

Windows:

- capable of being opened safely
- when open not project and cause a danger to passers-by.

r.16 Windows must be capable of being cleaned safely (possibly using special equipment)

r.17 Traffic routes:

- include both pedestrian and vehicular traffic
- pedestrian routes should:
  - wherever possible be separated from vehicular routes
  - be of adequate width
  - have non-slip surface
  - accommodate the disabled including wheelchairs
  - have wheelchair ramps where steps occur
  - be clearly identified by signs and/or floor marking
  - kept clear of obstructions
  - have protective guard rails at blind exits from buildings
- vehicular routes should:
  - be wide enough to accommodate the vehicles that use them
  - be of suitable construction to support the vehicles
  - be kept well maintained without potholes
  - be properly drained
  - have speed limits with speed humps
  - be clearly marked
  - have signs and road markings that comply with road traffic signs and markings
  - have hazards such as limited headroom or width clearly identified with advance warnings
  - avoid hazards such as the edge of pits or trenches, susceptible structures such as cast iron support columns, power cable poles, chemical filler pipes, etc.
  - allow for passing points and parking
  - have adequate manoeuvring space by loading bays
  - have safety refuges in loading bays
  - be provided with pedestrian crossing points.

r.18 Doors and gates:

- all doors and gates should be:
  - of proper construction for their function and size

- fitted with proper latches and closing devices
- if opening both ways have a viewing panel
- power operated doors and gates should:
  - have means to prevent anyone being trapped such as:
    * sensitive edge trip that causes the door to open
    * closing force low enough not to cause injury
    * manually operated control by hold-down button, release of which causes the door to open
    * if power fails:
      - be capable of being opened manually or
      - open automatically on power failure
      - except lift gates which must remain closed and only open at landings
  - have a local isolating switch for use in emergencies
- sliding doors should:
  - be fitted with a device, such as a fixed rail over the running wheels, to prevent them leaving the track
  - have means to guide the lower edge
- vertical rising doors should have an anti-fallback device, i.e. be counter-balanced or fitted with a ratchet
- fire doors are intended to retard the spread of fire and should:
  - be installed at fire exits and along fire escape routes
  - be capable of retarding fire for the time specified in the Fire Certificate:
    * most common is $\frac{1}{2}$ hr fire break-door
- fire doors on traffic routes through fire-break walls may be held open by a device such as a fusible link or magnetic device activated by the fire alarm that release the door automatically in the event of a fire
- smoke doors fitted in corridors should:
  - be of the $\frac{1}{2}$-hour fire-break type
  - close to a rebate or, if double swing, have edge seals to contain smoke
  - be self closing
  - be fitted with a viewing window.

r.19 Escalators and travelators should:

- be fitted with devices to prevent trapping between:
  - the treads and end-comb
  - treads and side-plates
- have readily accessible and easily identifiable emergency stops.

r.20 Toilets should be:

- easily accessible and clearly identified
- under cover
- separate for men and women
- private
- arranged so that urinals are not visible from outside men's toilet when door open
- fitted with obscured glass in the windows or curtained
- kept clean and tidy
- well ventilated and lit

- provided with a ventilated space between toilet and work/public area
- provided with facilities for washing hands and drying them.

WC cubicles should:

- have locks on the inside of the doors
- be large enough to give privacy
- be provided with toilet paper.

Number of toilet facilities to be provided:

- in post-1992 work premises:

max number on premises	no. of WCs	washbasins
1–5	1	1
6–25	2	2
26–50	3	3
51–75	4	4
76–100*	5	5

* thereafter one WC and one washbasin per twenty-five persons

  – if premises occupied by men only:

max number on premises	no. of WCs	no. of urinals
1–15	1	1
16–30	2	1
31–45	2	2
46–60	3	2
61–75	3	3
76–90	4	3
91–100*	4	4

* thereafter one WC and one urinal per fifty men

- in pre-1993 premises:
  – factories:
    *1 WC per 25 women employed
    *1 WC per 25 men employed (urinals are additional but no number given)
    *in factories employing over 100 men the ratio of facilities for men reduces
  – offices (for both sexes):

number employed	no. of WCs
1–15	1
16–30	2
31–50	3
51–75	4
76–100*	5

* thereafter one WC per twenty-five employees
  – if urinals are provided for men, the number of WCs reduces by one.

r.21 Washing facilities should:

- be located in or near toilet facilities and/or changing rooms
- be easily accessible

- be supplied with hot and cold water
- be provided with soap and towels or other method of drying
- be well ventilated and lit
- be kept clean and tidy
- have separate facilities for men and women unless washing is restricted to hands, arms and face only
- as a guide, have one wash basin per WC – if work is dirty there should be one washbasin per ten employees
- if showers are provided they should:
  - enjoy the same privacy as toilets
  - have thermostatic control on hot water supply.

r.22 Drinking water supply must be:

- adequate
- potable
- readily and easily accessible
- clearly marked and suitably positioned
- provided with 'Drinking water' or similar label on tap
- drinking fountain or provided with cups or drinking vessels.

r.23 Clothes accommodation:

- for storage of non-work clothes
- separate storage of work clothes
- clothes storage must be secure
- provide facilities for drying wet clothes including low-level heating.

r.24 Changing rooms should be:

- provided for changing clothes
- separate for men and women
- provided with:
  - benches or seats
  - storage facilities
  - heating
  - washing facilities if appropriate
- large enough to accommodate the number likely to be changing at any one time.

r.25 Rest rooms should:

- be available or a rest area
- be private
- have facilities for lying down
- have curtains or other means to subdue the light
- be adjacent to first aid facilities and toilets
- include facilities for pregnant women and nursing mothers to rest
- ban smoking
- be kept clean and tidy preferably under control of nurse or first aider.

> Eating facilities (small facilities and not canteens with paid staff) should:
>
> - be separate from work area
> - include the provision of hot water or hot and cold drinks dispensers
> - be provided with suitable tables and chairs
> - be provided with a sink or means for washing eating utensils
> - be kept clean and tidy (it may be subject to hygiene inspections)
> - contain a refrigerator for keeping foods fresh
> - have means for heating food – small oven or microwave
> - ban smoking (a separate room could be provided for smokers).

Any signs provided to indicate the location of facilities and escape routes should comply with the *Health and Safety (Safety Signs and Signals) Regulations 1996*.

These requirements extend considerably what was required under FA but only demand what is reasonable for a decent quality of working life.

5.2 Office safety

While offices are workplaces and as such have to comply with the requirements of the *Workplace (Health, Safety and Welfare) Regulations 1992* as summarised in Section 5.1 they do contain a number of hazards that are specific to them and for which there are particular precautions.

The following is a list of some of the hazards met in offices and typical precautions that can be taken.

Location	Hazard	Precautions
Floor	Polished and slippery	Use non-slip polishes Cover floor with non-slip mats
	Water or greasy spillages	Clean up as soon as found
	Threadbare carpets	Tripping hazard; remove or replace
	Man-made fibre carpets	Generate static electricity when walked on particularly with plastic-soled shoes Sprinkle carpet with water or use anti-static spray
Filing cabinets	Overloaded drawers	If top drawer is heavy, cabinet can topple when drawer opened Keep heavy loads to lower drawer Attach cabinet to wall or to back of other cabinet

		Drawers left open	Form tripping hazard and obstruction Ensure drawers shut when not in use
	Shelves	At high level	Reaching up to place items Provide suitable steps
	Electrical equipment	Loose leads across the floor	Remove or enclose in special-purpose ramped floor conduit
		Condition of leads, etc.	Regular inspection of plugs, sockets, leads and appliances by qualified electrician and replacement of faulty items
		Private equipment	Use of unofficial equipment, i.e. kettles, radios, drills, heaters, etc. should be prohibited
		Repair of equipment	Carried out by qualified electrician only Equipment must be isolated from electrical supply
	Photo-copiers	Clearing jams	Power must be switched off Usually happens automatically when cabinet door opened Follow maker's instructions Do not use pointed instruments to probe
		Fumes	Ozone may be generated (OES – 0.1 ppm) Ensure area is well ventilated
	Printing machines	In-running nips	Ensure operator fully trained Nip guard is in position
		Use of solvents and inks	Keep quantity to minimum Use non-spill containers Provide good ventilation Provide protective gloves Provide washing facilities nearby Mop up any spillages
	Fire precautions	Smoking	Prohibit Provide special designated smoking areas with facilities for extinguishing dog-ends
		Smoke and fire doors	Keep closed If volume of traffic requires them to be open, retain in open position by automatic device which releases when fire alarm activated

	Extinguishers	Correct type for possible type of fire
Regularly inspected and maintained
Properly mounted 1 m. from floor level
	Fire escape routes	Clearly marked
Kept clear at all times
Known to all staff
	Fire-exit doors	Must be kept unlocked at all times if persons on premises
If for security reasons must be locked, use panic bar or break-glass bolt
Wastepaper baskets	Contents	Restrict to paper only
Risk of fire from dog-ends; prohibit smoking
Other items such as broken glass or razor blades to be parcelled separately and put to one side for cleaner to collect
Metal or non-flammable construction
Display screen equipment (DSE or VDU)	Carry out risk assessment	See Section 3.4
	Work station	Ergonomic layout – see Section 10.3
	Noise from some printers	Provide acoustic hood
	Radiation emissions	Very low-level and no risk to health or pregnancy
	Rest pauses	Regular breaks from work station – move around, especially important for pregnant women
Furniture	Condition	Well maintained
Suitable for the work
Remove sharp corners and splinters
Ventilation	Natural ventilation	Openable windows
	Forced ventilation	Window or wall-mounted fans – blades guarded
Free-standing circulating fans – blades guarded

82 Workplace safety

	Draughts	Adjust means of ventilation to reduce to minimum
	Smokers	Ban
Lighting	Adequate level	Properly designed artificial lighting system
		Eliminate glare
		Use natural light wherever possible
		If glare from sunlight, provide suitable blinds

Although offices are generally considered to be safe they still contain many hazards that can cause serious injury. The majority of office accidents are of behavioural origin and can be avoided by taking care, being considerate for others and treating equipment with respect.

5.3 Workplace safety signs and signals

The use of the correct safety signs in the workplace can:

- re-inforce safety instructions and rules
- give information on risks and precautions to be taken.

The type, shape and colour of the signs are specified in *The Health and Safety (Safety Signs and Signals) Regulations 1996* which apply only to workplaces. Signs concerning the identification of hazardous and dangerous substances are dealt with in Sections 11.2, 11.4 and 11.8.

The Regulations require:

r.4 Where, after a risk assessment, residual risks remain suitable warning signs must be posted to indicate the nature of the risk and precautions to be taken.
  Suitable fire safety signs to be posted to meet requirements of a Fire Certificate or the recommendations of a Fire Authority (additional signs may be necessary on the recommendations of fire insurers).
  Any signs provided must be kept maintained and in place.
  Where a risk arises from road traffic, a sign prescribed by the *Road Traffic Regulation Act 1984* may be used if it is appropriate.
  Where hand signals are used they should comply with the signals illustrated in either:

- part IX of schedule (see 5.3.2 below) or
- BSS 6736: 1986 Hand signals for agricultural operations, or
- BSS 7121: 1989 Code of Practice for the safe use of cranes, or
- Appendix C of the Fire Training Manual

> r.5 Employers must provide for their employees:
> - comprehensive information on the interpretation of the signs
> - training in
>   - the meaning of the signs
>   - action they require to be taken.

There are four colours which have specific meanings:

5.3.1 Safety sign colours

Colour	Meaning	Information
Red	Prohibition sign	Action shown must NOT be carried out
	Danger sign	Shutdown, evacuate, emergency devices have operated, stop actions
	Fire fighting equipment	Identification of equipment and its location
Yellow	Warning sign	Take care, take precautions, proceed cautiously
Blue	Mandatory sign	Instruction MUST be followed. Equipment indicated MUST be worn
Green	Safety information sign	Emergency escape routes, location of first aid post

The signs associated with these colours are of a particular design and shape:

Prohibition sign
- round shape
- black pictogram on white ground
- red edging and diagonal line

84 Workplace safety

Fire fighting sign
- rectangular or square shape
- white pictogram on red ground

Warning sign
- triangular shape
- black pictogram on yellow ground
- black edging

Mandatory sign
- round shape
- white pictogram on blue ground

Workplace safety

Safety information sign
- rectangular or square shape
- white pictogram on green ground
- can be to BS 5499 pt 1: 1990 or to EU Directive 92/58EEC but must be consistent throughout workplace

Detailed information on the material and size of these signs can be obtained from reputable sign manufacturers who will also survey premises and advise on number and location of signs.

5.3.2 Safety signals

These are hand signals used in handling, lifting and other operations where the plant operator cannot see what is happening to the work piece.

In any such operation:

- there should be only one person (banksman) instructing the operator
- both the banksman and the operator should have the same understanding of the signals which should be to BS 7121: 1989
- the banksman should be clearly visible to the operator at all times (or in direct telephone/radio contact).

Typical hand signals include:

Stop   Emergency Stop   Clench and unclench fingers to signal 'take the strain' or 'inch the load'

Hoist   Lower   Slew In Direction Indicated

Workplace safety

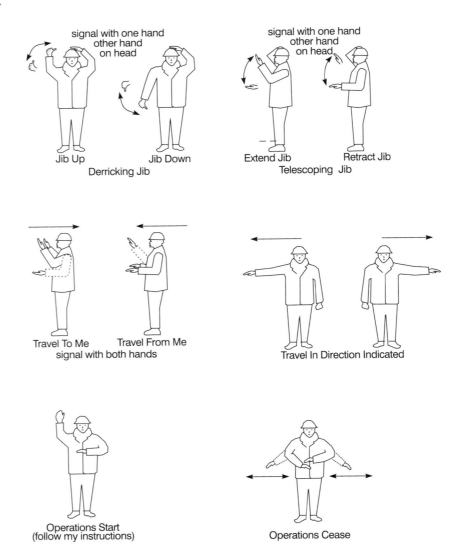

Safety signs and signals are there to inform and make work operations safer, but they will only be effective if the workpeople know what they mean and use them.

6 Information and advice

Information, with its attendant need for communication, is the lubricant that ensures the smooth running of an organization. This applies very much in health and safety where, without it the motivation and enthusiasm would very quickly dry up and safety activities grind to a halt.

But it is essential that the information, when provided, is well founded with a sound professional backing.

This chapter looks at sources of safety advice and information, the transfer of information by way of reports and how information can be stored so that it can be easily retrieved.

6.1 Safety advice

Under the *Management of Health and Safety at Work Regulations 1992*, all employers should have available to them a source of competent health and safety advice. How this is achieved will depend on the size and organization of the particular employment but it can be achieved in a number of ways.

Safety advice can be provided by:

- full-time safety adviser(s)
    - normally only where:
        * the concern is large enough to support them
        * the degree of risk warrants it
- part-time appointment
    - part of a dual role for an existing employee
    - appointment of a part-time consultant
- retained consultant
    - retained to provide:
        * an agreed number of days advice
        * advice on demand at an agreed fee.

The employer must ensure that the person appointed is competent in relation to the advice to be given.

Competence can be shown by:

- for a full-time safety adviser and a consultant:
    - being qualified in health and safety:
        * having a university degree in occupational health and safety, or
        * having passed Part II of the NEBOSH Diploma examination (NVQ level 4)
    - **and** having a number of years practical experience in the field
    - being a full (corporate) member of IOSH
    - being a Registered Safety Practitioner (RSP) with IOSH
    - being a registered MIIRSM with the British Safety Council (BSC).

- for a part-time appointment:
  - in high-risk premises:
    * being qualified to the same level as full time appointee
    * having passed Part 1 of the NEBOSH Diploma examination (NVQ level 3)
    * having some years practical health and safety experience
  - in lesser-risk areas
    * having passed the NEBOSH National General Certificate examination
    * being familiar with the work area
  - in low-risk premises
    * having passed the NEBOSH National General Certificate examination
    * having obtained the IOSH 'Managing Safely' certificate
    * being familiar with the work area.

Where a professional consultancy is selected:

- the employer should check that they have:
  - the necessary degree of knowledge
  - the necessary degree of expertise
  - sufficient facilities to provide the appropriate level of advice.

Whichever method is used the employer is required to:
- give information on particular hazards or factors involved in the work
- ensure that suitable facilities are available to carry out any necessary investigation
- allow adequate time for carrying out the safety function
- if more than one person appointed, to ensure they co-operate.

HSE inspectors will give advice if requested but you can never tell where advice stops and inspection begins.

The fact of appointing a safety adviser does not absolve employers from their statutory responsibilities – the role of the safety adviser is to give employers advice that will assist them in making decisions on how to meet their statutory obligations. However, it is common for employers to delegate the *performance* of their obligations to a manager or other appointee.

The employer is required to consult with safety representatives on the appointment of safety advisers.

6.2 Sources of information

In the course of their daily work, managers and safety advisers need to obtain information about a whole range of subjects. It is not realistic to expect them to have all the information they need at their fingertips but they should know where to go to get that information. This section looks at some of the sources of information available to managers and safety advisers.

It is convenient to consider this in two parts; written information and oral information.

6.2.1 Sources of written information

Laws:

- Acts (statutes), Regulations and Orders (statutory instruments)
- all part of statute law
- copies obtainable from HMSO

- copies held by public reference libraries
- catalogues of statutes held by reference libraries
- state the obligations to be met.

HSC/E* publications:

(Copies obtainable from HSE Books, PO Box 1999, Sudbury, Suffolk CO10 6FS)

- Approved Codes of Practice – COP series:
  - Codes of Practice approved by the Minister
  - set standards to give compliance but recognize that other methods may be equally effective
  - quasi-legal status and can be used in court to show how obligations can be met
- Health and Safety Regulations booklets – HS(R) series
  - give interpretations of regulations and how to comply
  - being superseded by Legal booklets, L series
- Health and Safety Legal booklets – L series
  - give guidance on the interpretation of and compliance with specific pieces of health and safety legislation
- Health and Safety Guidance booklets – HS(G) series
  - give guidance on techniques and practices for a range of activities and situations to achieve compliance with statutory requirements
  - no legal recognition but may be referred to by inspectors to indicate means of compliance
- Guidance Notes – five different series of guidance on specific subject topics:
  - Chemical Safety (CS series)
  - Environmental Hygiene (EH series)
  - General Series (GS series)
  - Medical Series (MS series)
  - Plant and Machinery (PM series)
- a range of free pamphlets giving advice on a wide range of occupational safety and health issues.

Further information on current health and safety issues, including developments in the EU, are given in HSC Newsletter published bi-monthly.

HSE also provide:

- a databased bibliographic reference via HSELINE on subscription.
- an information service via HSE Infoline – Tel: 0541 545500.

Other sources of published information include:

- British Standards Institution
  - publishes British Standards – identified by number BS XX
  - publishes EU Harmonised Standards in UK – identified by number BS EN XX

*The above HSC/E guidance publications are listed in the HSE's publication 'Price list of current HSE publications' obtainable free from HSE Books. This price list also contains information on IT services offered by HSE. The free pamphlets are listed in HSE's publication 'List of Free Publications' available free from HSE Books.

- EN take precedence over BS standards
- compliance with BS and BS EN standards deemed to give comformity with statutory requirement.

- Books
  - *Health and Safety* by Redgrave, Fife and Machin, Butterworth-Heinemann – the 'bible' of health and safety law
  - *Safety at Work* by John Ridley – 'authoritative guide to health and safety', Butterworth-Heinemann
  - *Croner's Health and Safety Manager*, loose-leaf information up-dated three times a year
  - *Tolley's Health and Safety at Work*, loose-leaf information with regular up-dating
  - *Health, Safety and Environment Bulletin*, Industrial Relations Services – tends to deal with the legal aspects, useful for case law and Tribunal decisions
  - *Hazards at Work*, TUC – useful book giving sound advice but tends to be from the employee's point of view.

- Commercial safety booklets such as:
  - Scriptographic Publications booklets – a range of booklets giving simple explanations, with cartoon illustrations, of various health and safety techniques and practices. Anglicized American but useful for issue to employees.

- Journals of the professional institutions:
  - The Safety Practitioner
  - IOSH publications.

6.2.2 Sources of oral information

Safety advisers are generally happy to discuss safety problems on the phone subject to there being no commercial implications. Employer/employee organizations and the national safety organizations tend to give advice only to their members. Safety consultants are likely to charge for any advice.

- Safety adviser
  - should be able to give information pertinent to the particular work situation

- Safety representative
  - can often give information about work hazards and solutions

- HSE inspector
  - able to give authoritative advice on a range of occupational safety related problems but if asked to call may use the opportunity to carry out an inspection of the premises

- employer's organization
  - many have a safety department able to give advice pertinent to the particular industry

- employee's organization
  - many have a safety section but may restrict advice to members; also advice may be strongly angled towards the employee

Information and advice

- The Royal Society for the Prevention of Accidents (RoSPA)
  - country's longest established safety organization
  - covers safety in all human activities
  - can give occupational safety advice but may restrict it to members

- The British Safety Council
  - an independant safety organization
  - gives advice but only to members

- Manufacturers and suppliers
  - have legal obligation to provide customers with safety information about their products
  - usually willing to assist with safety problems affecting their products

- Safety consultants
  - should be professionally qualified but credentials should be checked before using
  - can give detailed technical advice in respect of specific problems
  - can be very expensive.

6.3 Report writing

Reports are a two-edged sword – they can be used to get a message through to the recipient (manager) or be a means whereby the recipient finds things out about what is going on.

To be effective, reports must:

- be properly constructed
- be structured
- present the information in a logical manner
- be concise while covering the subject
- readable with good grammatical construction
- use acceptable words not jargon
- carry the reader to the desired conclusion.

The stages in preparing a report are:

- gather information and facts
- analyse the facts
- organize the data
- put pen to paper.

Points to be kept in mind when preparing a report:

- why is it needed?
- the object of the report
- the audience it is aimed at
- what it should cover and not cover
- whether it should be general or specific and detailed
- need for the text of the report to 'flow'
- possible contents
- how it is to be structured
- identification of headings and sections – numbering
- value of diagrams, graphs and illustrations.

Suggested make-up of a report:

- title page with author's name and signature, date of issue and distribution
- summary
- list of contents
- introduction giving background information on the subject
- description of the investigation/incident/circumstances
- findings – recording of information obtained
- analysis of findings
- conclusions drawn from findings
- recommendations
- appendices/annexes of supporting material such as photographs, sketches, diagrams, tables, etc.

Presentation of report:

- important to attract attention
- in acetates or special covers
- title displayed on or through cover
- binding – ring/comb/staple/perfect (glued)/sewn
- distribute with covering letter.

Before finalizing a report:

- read it through
- ask yourself:
  - does this add anything to the subject of the report? If not, discard it
  - is that what I meant to say? If not, discard or change it to say what you meant
  - can that be interpreted another way? If so, rewrite it
  - will that be clear to the reader? If doubtful, rewrite less ambiguously
  - is it to the point? If not, discard it.

Remember the recipient of the report is a busy person and to keep his/her attention follow the KISS principle – **K**eep **I**t **S**hort and **S**imple.

Useful references:

- *The complete plain words* by Sir Ernest Gowers, Penguin Books
- *A Guide to Report Writing* (booklet), the Industrial Society.

6.4 Data storage and retrieval

There is no sense in collecting data if you are not going to use it, and if you do collect it how are you going to store it so you can find it when you want it? Whatever system is used must match identified needs and integrate with existing data storage and retrieval systems.

6.4.1 Reasons for keeping health and safety data

- Statutory
  - accident reports
  - risk assessments required under various Regulations
  - reports of statutory inspections of pressure systems, lifting equipment, etc.
  - chemical data sheets under COSHH
  - training records
  - Fire Certificate (if issued).

Information and advice

- For use in company
  - reports of accident investigations
  - reports of safety inspections
  - minutes of safety committee meetings
  - information on health and safety items and products
  - copies of health and safety laws
  - personnel records
  - risk assessments.

6.4.2 Methods of storing data

- By diary
  - simple risk assessments
  - notes of minor and non-reportable accident (sometimes in the Accident Book form BI 510).

- By hard copy
  - reports of statutory examinations
  - chemical data sheets
  - copies of accident report forms F 2508 and F 2508A
  - HSE Codes of Practice, guidance notes and advisory publications
  - copies of Acts and Regulations.

- On disc
  - scheme for examination of pressure systems
  - safety information (HSE's Infoline: HSELINE; etc.)
  - accident data and analysis.

- Microfiche
  - safety information.

6.4.3 Information retrieval

- From diary
  - by memory.

- Hard copy
  - from particular file.

- From disc
  - by using search or interrogation facility.

6.4.4 Systems for identifying stored data

- Hard copy
  - subject matter
  - material
  - numerical referencing system
  - alphabetic referencing system
  - haphazard system (memory).

- Disc
  - subject heading
  - keywork.

This may look like a lot of paper or office work but it is necessary if benefit is to be gained from past experiences and also, in certain cases, as proof of actions taken should queries arise in the future.

7 Accidents

Accidents occur – they do not *happen*; they are *caused* – by short-comings on the part of the employer, the employee or both. The results can be traumatic for both: for the employee in the way an injury can affect him personally, his family and their quality of life; for the employer through lost production, time spent on investigations and at worst, the cost of a legal action.

Accident prevention aims to reduce the chances of such an incident to the absolute minimum.

7.1 Principles of accident prevention

Objectives

1. To prevent accidents happening.
2. If accidents do occur, to prevent recurrence.

Procedure

- Identify hazards.
- Eliminate hazards.
- If hazards cannot be eliminated, reduce hazard to minimum.
- Assess residual risk.
- Control residual risk.

Identification of potential hazards

- Pre-accident
  - risk assessment (see Section 3.4)
  - safety inspections
- Post-accident
  - accident investigation (see Section 7.2)
- Near miss
  - implement near-miss reporting procedure.

Definitions

Hazard – something with the potential to cause harm
Risk – the probability that the harm will occur and the severity of the resulting damage
Accident – an unforeseen event causing injury or damage
Near miss – an event that narrowly missed causing injury or damage.

7.1.1 Heinrich's domino theory

An accident is not a single event; it is the result of a series of linked causes.

The dominoes in Figure 7.1 represent the sequence of causes (events or situations) leading to an accident that results in injury or damage. If a domino falls it will knock over the rest until eventually the final domino

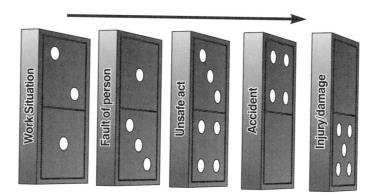

Figure 7.1 Heinrich's domino theory

falls, i.e. an accident happens. If any of those dominoes (causes) is missing, i.e. the correct safety action has been taken, there will be no accident.

Typical examples of these causes are:

- Work situation
  - inadequate management control
  - lack of suitable standards
  - failure to comply with standards
  - faulty or inadequate equipment

- Fault of person
  - lack of skill or knowledge
  - physical or mental problems
  - lack of or misplaced motivation
  - inattention

- Unsafe act
  - not following agreed methods of work
  - taking short cuts
  - removing or not using safety equipment

- Accident
  - the unexpected event
  - contact with dangerous machinery or electricity
  - falls
  - being struck by equipment or falling materials, etc.

Resulting in:

- Injury/damage
  - to employee
    * pain and suffering
    * loss of earnings
    * loss of quality of life
  - to employer
    * damage to plant
    * compensation payments
    * loss of production
    * possible prosecution.

96 Accidents

The accident triangle (from HS(G)65)

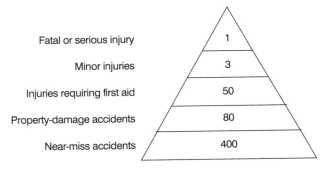

Fatal or serious injury	1
Minor injuries	3
Injuries requiring first aid	50
Property-damage accidents	80
Near-miss accidents	400

7.1.2 Practical accident-prevention techniques

- Near misses
  - encourage the reporting of near misses
  - investigating them can prevent serious injuries

- Hazard identification
  - by inspection
  - through safety tours, inspections, etc.
  - reports from operators

- Hazard elimination
  - by engineering means
  - changing plant
  - changing materials
  - changing the process

- Reduction of hazard
  - engineering means
  - provision of guards
  - provision of PPE

- Assessment of residual risk
  - see Section 3.4

- Control of residual risk
  - engineering means – alarms, trips, etc.
  - safe systems of work
  - training of employees.

7.1.3 Management techniques

- Commitment
  - implement incident-reporting procedure and monitor it
  - manager seen to be involved in and committed to preventing accidents
  - effective safety policy
  - establishment of suitable formal organization with stated safety responsibilities
  - development of appropriate rules and standards
  - ensuring good communications
  - have effective consultative procedures/committees
  - use of job-safety analysis

- monitor performance and follow up short-comings
- include safety from design stage
- quality assurance and safety – BS ISO 9001 registration.

7.1.4 Role of engineering

- Knowledge
  - need to know techniques for guarding machinery
  - appreciate operator working methods

- Maintenance
  - high-risk area
  - need for operatives to be skilled
  - technical training
  - safe systems of work/permits-to-work

- Plant and equipment
  - maintain in good repair
  - preventative maintenance.

7.1.5 Costs of failure

- Direct costs
  - sick pay
  - repairs to damaged plant
  - loss of production
  - increased insurance costs

- Indirect costs
  - cost of investigation
  - loss of goodwill and/or image in the community
  - hiring and training replacement staff.

7.2 Accident investigation

Objectives

- To determine the cause so recurrence can be prevented.
- Not to apportion blame.
- To obtain information for reporting to enforcing authority.
- To obtain information for insurers either:
  - in respect of a claim for pain and suffering by injured person to assist settlement or to resist litigation
  - to claim for damage to plant, equipment, etc.
- To obtain information for other statutory reporting, i.e. Social Security benefits.

Cause of accident – the events or circumstances that led to the incident that caused the injury or damage.

- Immediate cause – the part or item that actually caused the injury or damage.
- Root cause – the action or activity that resulted in contact with the immediate cause. Root-cause analysis involves checking on the sequence of events and decisions that led to the accident and identifying the often remote action that triggered that sequence of events.

Cause of injury or damage – the action or process that occasioned the actual injury or damage.

Investigation

- By whom?
    - initially supervisor, who informs safety adviser
    - safety representative – note their rights (see Section 4.3)
    - safety adviser
    - if a claim against employer likely or has been made, insurance surveyor/engineer
    - if the injury or incident has to be reported to enforcing authority, enforcing authority inspector
    - if a fatality, police.

- When?
    - immediately injured person returns from first aid or is removed for treatment
    - before site of accident has been disturbed.

- Procedure
    - view site and note significant details
    - take photographs
    - measure up relevant parts and areas
    - check condition of plant and equipment – arrange for tests if necessary
    - interview witnesses
        * ideally alone but may have representative present if requested
        * point out that the object of investigation is to discover the cause of accident
        * evidence must be direct not hearsay
    - check records of training given to injured employee
    - interview injured person as soon as possible but do not cause distress
    - analyse information and prepare report
    - if claim has been entered, insurers agent will want to investigate and interview witnesses but cannot interview claimant
    - if enforcing authority inspector investigates, may take statements from witnesses including injured person
    - in case of fatality, police investigate to determine cause of death and if there has been any criminal action.

- Inquiry
    - if inquiry held, purpose must be clearly stated, i.e. to determine the cause of the accident
    - report of inquiry available to both employer and employee, i.e. is 'discoverable' in the event of a claim
    - if object of inquiry is to resist a claim, this must be clearly stated and understood by all those involved, when notes and report could be 'priveleged'.

- Information to collect
    - details of site – owner, address, department/section/shop
    - process or operation concerned including details of any plant involved
    - date, time of accident
    - personal details of injured person (possibly from personnel records)
    - information on training given to injured person

- job being carried out at time of accident
  * was it authorized?
  * was correct procedure being followed?
  * were guards in place?
  * etc.
- details of injury received.

- Report
  - see Section 6.3
  - analyse results of investigation and information obtained
  - prepare report setting out circumstances of accident and possible causes
  - make recommendations to prevent recurrence.

7.3 Accident reporting

There are three reasons for reporting accidents

1 Statutory obligation
2 Insurance claim
3 In-house accident prevention.

7.3.1 Statutory reporting

This is required by the *Reporting of Injuries, Diseases and Dangerous Occurrences Regulations 1995*, which imposes requirements relating to the occurrence of:

- fatalities
- major injuries
- injury to other than employee requiring removal to hospital
- certain diseases caused by listed processes
- injury or death associated with the use of flammable gas
- one of the listed dangerous occurrences.

Injuries to be reported

- Major injuries:
  - any fracture other than to finger, thumb or toes
  - any amputation
  - loss of sight – whether temporary or permanent
  - a chemical or hot metal burn to the eye or any penetrating injury to the eye
  - electrical injuries such as burns, shock and loss of consciousness
  - any other injury causing hypothermia, heat-induced illness or unconsciousness, requiring resuscitation or admittance to hospital for more than twenty-four hours
  - loss of consciousness due to lack of oxygen
  - decompression sickness requiring immediate medical attention
  - acute illness needing medical attention or loss of consciousness from a chemical substance
  - acute illness from pathogen or infected material
  - any injury that results in detention in hospital for more than twenty-four hours.

- When?

- major injuries
  forthwith by quickest possible means and follow with a written report within ten days
- diseases
  on receipt of written diagnosis from a doctor of one of the diseases listed in the Regulations
- flammable gas incident
  on receipt of notification of an injury from flammable gas followed by a written report within fourteen days
- dangerous occurrence
  forthwith by quickest means and follow with a written report within ten days.

- How?
  - for injuries, gas incidents and dangerous occurrences
    form 2508
  - for diseases
    form 2508A.

- By whom?

- employer
- person in control of the work
- the manager of a mine
- the owner of a quarry
- the owner of a tip
- the owner of a pipe line
- if from materials being carried, the owner of the lorry.

Statutory records and accident reports must be kept readily available for inspection and retained for a period of at least three years.

Under the social security legislation, the employer is required to have and keep available a copy of Form BI 510, the 'accident book', for accidents to be noted in if the employee requests it. This book is sometimes used as a convenient means for recording all injuries.

Note: benefit under the current Social Security Act may be claimed for certain diseases and injuries. These claims are initiated by the claimant but the employer may be required to confirm the circumstances of the incident giving rise to the condition:

- for injuries – form BI 76
- for diseases – form BI 76D.

7.3.2 Reporting to insurers

Where, following an injury-causing accident, it appears likely that the injured person will claim damages, employers may wish to make a claim under their Employer's Liability insurance policy. Insurers will want details of the accident and will send their own agent to investigate the incident. The agent will be acting on behalf of the employer and should be given full co-operation. From this investigation the insurers will assess the likely damages and make suitable financial provisions and also use the information to decide whether to negotiate a settlement or to resist the claim if it is made.

7.3.3
In-house reporting

This has three functions:

1. It lets the senior managers know an accident has occurred and the facts of its circumstances.
2. It may indicate a short-coming in safety arrangements within the organization and enable these to be corrected.
3. It provides the basic data from which domestic accident performance may be calculated.

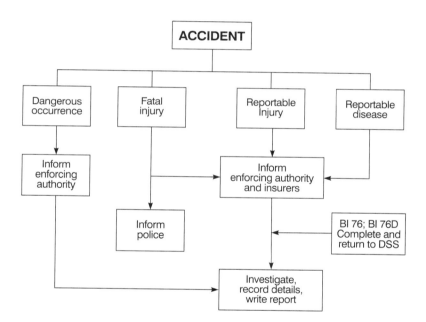

Figure 7.2 Diagram of action to be taken on learning of an accident

Part 3 Occupational health

Good health is an essential part of enjoying a good quality of life at home and at work. It is also an important factor in the viability of an organization. This fact is recognized by the Health and Safety Executive who have launched an initiative titled 'Good Health is Good Business'.

Many work situations, by virtue of the layout of the working station or of the materials used, present a higher than normal risk to health. By understanding the characteristic of the materials in use and the likely reactions of the body the risks to health can be reduced to a minimum. In understanding the body's reaction to chemicals it is necessary to understand how the body functions.

Part 3 deals with these aspects and with the on-site actions that can be taken should ill-health or an injury be suffered.

8 The body

Appreciating how the body works, how it reacts to a range of substances used at work and being aware of the ways in which those substances can enter the body are important aspects of reducing to a minimum the causes of ill-health to which employees may be exposed.

This chapter looks at those particular aspects of ensuring good health at work.

8.1 Functions of the body

The body is a complex organism comprising a great number of organs contained within a rigid structure (the skeleton) and held in place by various muscles. The different organs are all interdependent and play a specific role in the effective functioning of the body as a whole. However, the effectiveness of any of these organs can be adversely affected by conditions and substances met at work (and at home). The functions of some of the main organs and how they can be adversely affected are described below.

Organ	Function	Vulnerability
Bones	linked together they form the skeleton	they are brittle and can be broken by impact (blows) or sometimes by a muscular spasm
	red blood corpuscles are created in the bone marrow	this process is interfered with by toxic chemicals such as benzene and carbon monoxide and radioactivity
Skin	the protective layer that covers the outer surface of the body	can be penetrated by sharp objects and severe physical impact
		its protective fats can be dissolved by solvents resulting in dermatitis
		vulnerable to radiations
Muscles and tendons	control the movement of the various limbs	can be damaged by: – excessive loads (strains) – sudden jerks (sprains) – repetitive action (RSI and WRULD)
The gut	the digestive organs	means of changing ingested matter into substances needed by the body systems
		can be damaged by ingested corrosive and poisonous substances

106 The body

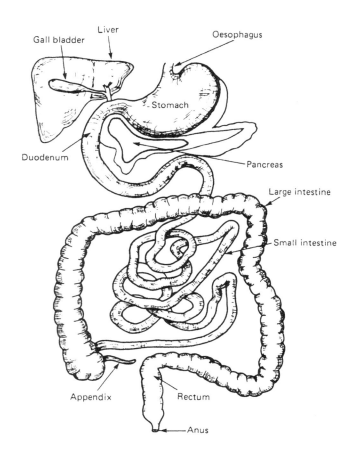

Figure 8.1 Diagram of digestive system

Liver	breaks down proteins from the gut, detoxifies body poisons and removes old red blood corpuscles	damaged by poisons such as organic solvents, certain metals, VCM and excessive alcohol
Kidneys	separates water and urea from body fluids and disposes of them	damaged by halogenated solvents and some heavy metals
Bladder	storage vessel for waste body fluids	prone to cancer from 2-naphthylamine
Lungs	the organ that takes oxygen from the air and transfers it to the blood	vulnerable to any respirable and inhalable fumes and dusts, particularly: – cancer from asbestos, radon and nickel – fibrosis from coal and silica dusts
Nerves	sensing organs that transmit messages to the brain	functioning impaired by exposure to toxic substances

The body **107**

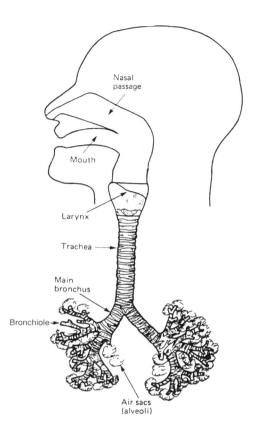

Figure 8.2 Respiratory system

Brain	the control centre for the whole body	narcotic effects of chlorinated solvents
		damaged by certain metals, carbon disulphide and carbon monoxide
Eye	the sight organ, delicate and exposed	vulnerable to: – dusts – flying particles – corrosive chemicals

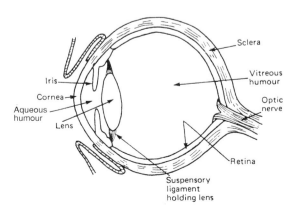

Figure 8.3 Diagram of section through eyeball

Ear	the hearing organ incorporating the organ for balance	hearing acuity can be damaged permanently by prolonged exposure to high noise levels
Nose	the organ of smell	very sensitive olfactory nerves desensitized by hydrogen sulphide
Heart	the body's pump supplies blood and oxygen to the brain, various organs and muscles	its muscles can be affected by electric shock, resulting in acceleration or stopping (fibrillation) of its pumping action.

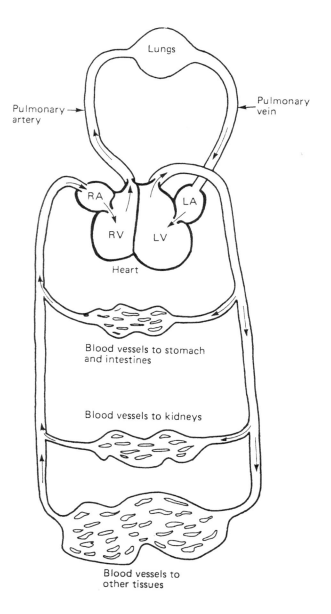

Figure 8.4 Diagram showing circulation of blood (RA – right auricle; LA – left auricle; RV – right ventricle; LV – left ventricle)

8.2 Routes of entry

Hazardous and dangerous substances are able to enter the body by a number of different routes:

- Ingestion through the mouth and into the gut.
- Inhalation into the lungs.
- Absorption through the skin.
- Direct entry through open cuts and wounds.

Simple precautions can prevent these entries:

- Ingestion
  - no eating in the work place
  - personal hygiene, washing hands before eating
  - not smoking in the workplace

- Inhalation
  - wear respiratory protection suitable for the particular substance
  - exhaust ventilation
  - fume and dust extraction

Target organs

- Absorption
    - wear protective gloves
    - wash in soapy water all contaminated areas
    - use a barrier cream

- Direct entry
    - get all cuts and wounds treated
    - keep all cuts and wounds covered when at work.

8.3 Target organs

Hazardous and dangerous substances do not attack the body indiscriminately. Different substances affect different organs although any particular substance may attack more than one organ. The hazards from substances can, broadly, be divided into seven groups and these are shown in the table below.

Type of hazard	Target organ	Reaction/symptom
Toxic	kidney, liver, bone marrow	attacks and affects the functioning of the organ
Carcinogenic	lung, liver, bladder	warts, ulcers, malignant growths
Corrosive	skin, lungs, stomach	destroys living tissues
Dermatitic	skin	inflammation of skin (dermatitis)
Irritant	skin, eyes, lungs	inflammation, dermatitis, fibrosis of lungs
Flammable	skin, whole body	burns
Radioactive	skin, sensitive organs such as bone marrow, eyes, gonads, etc.	leukaemia, cataracts, loss of fertility

Where a substance is used that affects more than one organ, it may be necessary to take separate precautions to protect each susceptible organ.

The hazard is related to the chemical characteristics of a particular substance.

Hazardous substances may occur in any of a variety of forms including solid, dust (particulate), gas, fume, liquid or a vapour (see Sections 11.1 to 11.7).

9 Health at work

Many substances in a variety of forms can have adverse effects on the health of employees. Having an understanding of the substances and the forms that present hazards to health can take you half-way in eliminating or reducing the risks.

The effects of some of the more common forms of substances are considered in this chapter.

9.1 Causes of health hazards

Where, at work, an employee is subjected to certain conditions or materials there could be a reaction resulting in ill-health. In many of these cases the hazards are known and the symptoms recognized. This section considers some of the main causes of occupational ill health.

Condition/material	Body reaction

Dusts	• when inhaled, affects the lungs causing pneumoconiosis. Particular dusts give rise to specific diseases:
	  – asbestos → asbestosis, mesothelioma
	  – silica → silicosis
	  – coal dust → pneumoconiosis

Poisons	• ingested poisons:
	  – can affect any one of a number of organs
	  – the body absorbs a certain amount of the poison very quickly
	  – the rest passes through and is evacuated
	  – vomiting should *not* be induced since it can cause more damage than the original poison

Solvents	• enter the body by:
	  – ingestion of liquids
	  – inhalation of fumes
	  – absorption through the skin
	• can have:
	  – narcotic effects on the nervous system
	  – toxic effects on organs such as the liver, kidney or bone marrow
	  – irritant effect by de-fatting the skin

Corrosives	• acids and alkalis • destroy living tissues • dilute by deluging with water • alkalis tend to be more persistent than acids • in both cases medical assistance should be sought

Condition/material	Body reaction
Irritants	• in the form of dusts or liquids may react with the skin and cause dermatitis • dusts when inhaled can cause irritation and fibrosis of the lungs

Carcinogens	• cause cancer in various organs including:
	  – the lungs through inhaling asbestos fibres
	  – the liver (as angiosarcoma) from VCM
	  – the skin from tar pitch
	  – the scrotum from mineral oils where there is poor personal hygiene
	  – the bladder by 2-naphthylamine

Gassing	• by the poisonous nature of the gas or fumes inhaled, i.e. chlorine, carbon monoxide, hydrogen sulphide, etc.
Asphyxia	• oxygen deficiency particularly due to heavier-than-air gases such as methane and carbon dioxide
Sensitization	• individuals can become sensitized to substances such as isocyanates, wood and flour dusts, moulds from rotting hay, colophony fumes from soldering, etc. • after sensitization, exposure to even the minutest concentration will cause a reaction • can seriously affect future working prospects especially where that particular substance is used

Bronchial asthma	• a prescribed industrial disease
	• caused by a range of materials including:
	  – isocyanates
	  – colophony fumes
	  – epoxy-resin curing agents
	  – wood and flour dusts
	  – animals and insects in laboratories

Diseases from metals	• diseases of varying seriousness can be caused by exposure to metals such as:
	  – lead
	  – mercury
	  – chromium
	  – arsenic
	  – manganese
	  – nickel
	  – cadmium
	  – vanadium

Condition/material	Body reaction

Ionizing radiations	- emitted by radioactive materials
	- ill-health effects range from:
	  – damage to sperms and white blood cells
	  – to nausea, vomiting, coma and death
	- strict controls must be exercised where radioactive substances are used (see Section 9.2)

Vibrating tools	- cause a variety of injuries to hands and arms – known generically as 'hand-arm vibration syndrome' (HAVS) - can cause constriction of the blood vessels in the hands, leading to a condition known as **vibration white finger** where the fingers go white and loose feeling - operators with this condition should be moved to other work
Noise	- major effect is **noise induced hearing loss** (see Section 12.1) - excessive noise can cause fatigue and disorientation

Heat and humidity	- work at high temperatures and high humidity can result in:
	  – cramps
	  – heat stroke
	  – exhaustion
	- no standards to work to but the chill effect of moving air can help

Micro-organisms	- include a number of organisms:
	  – viruses → hepatitis A and B, AIDS
	  – bacteria → anthrax, legionella, leptospirosis, tetanus, etc.
	  – fungi → Farmer's lung, ringworm
	  – protozoa → malaria
	  – nematodes → hookworm

Repetitive actions	- repeated forceful actions of upper limbs can cause:
	  – tenosynovitis
	  – carpal tunnel syndrome
	  – writer's cramp
	- sometimes referred to as repetitive strain injuries (RSI) or work related upper limb disorders (WRULD)
	- in early stages rest and removal to other work can result in recovery

Stress	- psychological reaction to factors often outside the individual's control such as:
	  – demands above or below ability
	  – the working environment
	  – relations with fellow employees or the organization.

114 Health at work

A nasty case of WRULD

Although there are many causes of health hazards at work, there are also a wide range of equipment and techniques to protect employees from them. Protection from these hazards is required by COSHH Regulations and is dealt with in Sections 11.1 to 11.8.

9.2 Ionizing radiations

Radiations from radioactive decay cause the material they pass through to become ionized, hence the term 'ionizing radiations'. The effect on body tissues depends on:

- the nature or type of radiation
- the dose and duration of exposure
- whether the source is internal or external to the body.

9.2.1 Types of radiation

- **α (alpha)** – a particle whose radiation is stopped by a few centimetres of air, a sheet of paper or the outer layer of skin. However ingested, α particles can cause intense local radiation and immense damage to the affected tissues.
- **β (beta)** – have greater penetrating power than an α particle but the ionization caused is less severe.
- **γ (gamma)** – electromagnetic radiation with great penetrating power. Caused by radioactive decay and emits radiation all the time.
- **X-rays** – electromagnetic radiations whose penetrating powers depend on its energy. Commonly created in X-ray machines, the radiations cease when the machine is switched off.
- **Neutrons** – neutrons are emitted during nuclear fission and have very great penetrating powers. They can cause intense ionization.
- **Bremsstrahlung** – electromagnetic radiations produced by the slowing down of a β particle. They can have considerable penetrating powers.

9.2.2 Effects on the body

These are determined by the dose received, i.e. the type and intensity of the radiations and the period of exposure. Special instruments – ionization chambers or Gieger-Müller tubes – are used to measure the dose rate. Exposure levels can be determined by the use of *film badges*.

- Small localized exposure can cause:
  - redness of the skin
  - cataracts in the eyes
  - loss of fertility.

- General whole body exposure can result in:
  - nausea, vomiting and diarrhoea
  - cancer of the skin and other organs
  - leukaemia.

9.2.3 Uses

- Low power sources (α and β), usually encapsulated or sealed, are used for:
  - thickness gauges
  - smoke alarms
  - static-electricity eliminators.

- High power sources (γ, X-rays and neutrons) are used for:
  - medical diagnostic investigations
  - non-destructive testing of materials
  - high-technology production processes.

9.2.4 Precautions

- Low power sources:
  - follow suppliers instructions
  - appoint a Radiation Protection Supervisor (an employee who has been suitably trained) to oversee the safe use of the sources
  - when not in use, keep the sources in a secure store.

- High power sources:
  - appoint a Radiation Protection Adviser who must be suitably qualified and experienced. This expertise can be bought in from the National Radiation Protection Board (NRPB) who provide this service as do a number of specialist consultancies
  - when not in use ensure the source is kept in a suitable shielded container in a secure store
  - when in use ensure that all employees except radiation specialists are clear of any area likely to be exposed to radiations
  - use sources only within properly shielded areas.

9.2.5 Legislation

The Ionising Radiations Regulations 1985 lay down the protective measures to be taken. These include:

- Restriction of employee's exposure.
- Control of access to areas where radiations may be present.
- Appointment of suitably trained or qualified persons to ensure safe use of sources.
- Implementation of rules for the safe use of sources.
- Training and instructing any employee who uses radiation sources.
- Measuring the exposure levels of employees working with radiations.
- Provision of medical examinations.
- Keeping accurate records of the use and locations of all sources.
- Reporting to the HSE any damage to or loss of a source.
- Investigating cases of over-exposure and taking corrective action.

9.3 Hazards from non-ionizing radiations

For detailed information on these requirements consult the Regulations, the NRPB or other suitably qualified consultant.

Where the use of radiation sources is contemplated, advice should be sought from the supplier, NRPB or suitably qualified consultant.

Non-ionizing radiations have a longer wavelength than ionizing radiations and do not cause ionization. However, they can give rise to serious risks to health and suitable precautions should be taken when they are used. They occur both naturally and in industry where they have a number of important uses.

Table 9.1 summarizes the types of non-ionizing radiation, their uses, hazards and precautions to be taken.

Table 9.1 Summary of non-ionizing radiations, uses and precautions

Type of radiation	Occurrence	Uses	Hazards	Precautions
Ultraviolet	Solar radiations Ultraviolet emitting equipment Welding arc flash	Sun beds Curing inks and paints	Erythema (sun burn) Arc eye Cataracts Skin cancer	Keep exposed parts covered Wear protective goggles Wear protective glasses Use barrier cream Keep equipment that generates ultra-violet fully enclosed with interlocks
Visible light	Solar radiations Lighting equipment	Illumination	Possible retinal damage from badly positioned luminaires	Control intensity of lighting and positioning of luminaires
Infra-red	Any high temp. surfaces Special infra-red generators	Process heating Heat-seeking devices Security and safety devices	Burns Cataracts	Provide guards Use insulated gloves
Radio frequency	Radio transmitters Radar Dielectric heaters Broadcasting equipment Microwave heaters	Communications Navigation systems PVC welding Cooking	Whole and partial body heating RF burns Cataracts	Provide screens Earth metal parts Check for radiation leakages
Electro-magnetic	Overhead power lines Induction heaters Zonal smelters Display screen equipment	Power transmission Domestic appliances	Nervous system dysfunction Irritation of facial skin	Move away Earthing appliances Very low risk, try earthing screen
Lasers	Special generating equipment	Illuminated displays Lining up plant and buildings Cutting wood and metals Entertainment Pointers	Mono-chromate sources over very wide range of wavelengths Delivers intense power to very small area Can cause corneal and retinal damage and skin burns	Enclosure of equipment, safety interlocks and shutters, wear goggles and gloves Competent trained operators Display warning signs

10   Health protection

There are a number of well established techniques for looking after the health of employees. These include taking action to prevent ill-health, providing means to prevent employees coming into contact with hazardous substances and ensuring that if they are hurt in any way their injuries are properly treated.

This chapter looks at the means to protect employees' health.

10.1 First aid

First aid is defined as:

- emergency treatment until the arrival of medical practitioner or nurse
- treatment of minor injuries that do not warrant medical attention or would otherwise receive no treatment.

First aiders trained by the Red Cross or St. John's Ambulance Brigade are no longer recognized unless their trainer had been approved by the HSE.

The first-aid facilities to be provided are outlined in the *Health and Safety (First Aid) Regulations 1981*, with greater detail being given in an Approved Code of Practice and Guidance 'First aid at work', HSE publication L 74. The recommendations include:

- The extent of the facilities depend on the risks faced, i.e. the higher the risk, the more extensive the facilities.
- There should be an adequate number of first aiders – one per fifty employees for low-risk work, with the number of employees per suitable person reducing as the risks increase.
- There should be a first-aid room if:
  - site is high-risk
  - site is remote from a hospital, e.g. in country areas
  - access to a hospital or doctor is difficult, e.g. in areas with bad traffic congestion
  - the number employed on the site warrant it.
- Where employees work away from base:
  - if work area is low-risk – no facilities
  - if work area is on another employer's premises – use local facilities
  - if area of work is high-risk or does not have access to first-aid facilities – a first-aid kit to be carried. (For contents see below).
- First-aid boxes must:
  - be constructed to protect contents
  - be kept replenished
  - contain a first-aid guidance card
  - not be used for keeping anything other than first-aid items.

118 Health protection

. . . a suitable person . . .

- Where more than one employer occupies a building or site, common shared facilities can be provided.
- Employees should be informed of first-aid facilities and their location.
- First-aid facilities may be made available to visitors, contractors, etc. on premises.
- Where a first-aid room is provided it must:
  - be under the control of a *suitable person* or nurse
  - have a *suitable person* available at all times anyone is working on the premises
  - where the *suitable person* is temporarily absent, have a responsible *appointed person* to deal with any first-aid treatment needed
  - be conveniently situated for ambulance access
  - be large enough to accommodate a couch
  - have doorways wide enough to allow passage of a wheelchair
  - be decorated with surfaces that can be cleaned easily
  - have hot and cold water for washing
  - be clearly identified
  - give location of nearest *suitable person*
  - be provided with a treatment book – this may be form BI 150 ('Accident Book') or can be the employer's own day book to record treatments.
- *Suitable persons* must:
  - be trained on an HSE-approved course
  - have received specific training where special hazards arise
  - record all treatments given
  - receive regular refresher training.

10.1.1 First-aid boxes

First-aid boxes should contain:

- a guidance card
- 20 individually wrapped sterile adhesive dressings
- 2 sterile eye pads with retaining attachments

- 6 individually wrapped triangular bandages
- 6 safety pins
- 6 medium-sized individually wrapped sterile unmedicated wound dressings
- 2 large individually wrapped sterile unmedicated wound dressings
- 3 extra-large individually wrapped sterile unmedicated wound dressings.

A supply of tap or sterile bottled water for irrigating eyes should be available. Additional items that can be provided include:

- a stretcher or other means of carrying a patient
- a pair of blunt-nosed stainless steel scissors
- a disposable plastic apron and gloves
- blankets
- a suitable bin for disposal of used swabs and dressings.

Travelling first-aid kits to contain:

- a first-aid guidance card
- 6 individually wrapped sterile adhesive dressings
- 1 large sterile unmedicated dressing
- 2 triangular bandages
- 2 safety pins
- individually wrapped moist cleansing wipes.

Employers are given a certain amount of discretion in the facilities they provide. First-aid box suppliers can be helpful but watch the pushy salesman.

10.2 Personal protective equipment

General requirements to provide personal protective equipment (PPE) are contained in the *Personal Protective Equipment at Work Regulations 1992*. However, there are specific requirements, which take precedence over these general requirements, which are contained in regulations dealing with particular hazards, namely:

> *The Control of Lead at Work Regulations 1980*
> *The Ionising Radiations Regulations 1985*
> *The Control of Asbestos at Work Regulations 1985*
> *The Noise at Work Regulations 1989*
> *The Construction (Head Protection) Regulations 1989.*

These latter Regulations are considered separately elsewhere and are not covered in this section.

In providing protection against a hazard, an employer's first priority must be to protect the workforce as a whole rather than individuals. The use of PPE should only be considered if more global methods of protection are not reasonably practicable or feasible.

With all PPE, the supplier will advise on the most appropriate type to provide the protection needed, and may offer a choice of material, design, colour, etc. However, there are some general principles that should be followed.

120 Health protection

... appropriate for the hazard faced ...

To be effective PPE must:

- be appropriate for the hazard faced
- be of a material that will resist that hazard
- be suitable for the person using it
- not interfere with the operator performing his/her functions
- be of robust construction
- not interfere with other PPE being worn at the same time
- not increase the risks to the wearer.

PPE should be:

- provided free of charge
- personal issue or given suitable hygiene treatment between uses
- used only for its intended purposes
- kept in good repair
- repaired or replaced if damaged
- kept in suitable accommodation when not in use.

Operators who use PPE should be:

- informed of the hazards faced
- instructed in the preventative measures taken

Health protection

- trained in the proper use of the equipment
- consulted and allowed a choice of PPE subject to its suitability
- taught how to maintain the equipment and keep it in good order
- instructed to report any deterioration or damage.

Examples of the protection provided by a range of types of PPE are:

Part of the body	Hazard	PPE
Head	falling objects	hard hats
	confined spaces	bump caps
	hair entanglement	caps, hair net, hair cut
Hearing	excessive noise	ear muffs, ear plugs
Eyes	dust and grit, flying particles	goggles, face shields
	radiations, lasers, arc welding	special goggles
Lungs	dusts	face masks, respirators
	fumes	respirator with absorbent filter (limited effectiveness)
	toxic gases and oxygen-deficient atmosphere	breathing apparatus
Hand	sharp edges and burrs	protective gloves
	corrosive chemicals	resistant gloves
	low/high temperatures	insulating gloves
Feet	slipping, sharp items on the floor, falling objects, liquid-metal splashes	safety shoes / gaiters and safety shoes
Skin	dirt and mild corrosives	barrier creams
	strong corrosives and solvents	impermeable barriers such as gloves and apron
Trunk and body	solvents, moisture, etc.	aprons, overalls
Whole body	hostile atmosphere (toxic fumes/ radioactive dusts)	pressurized suits
	falls	safety harness
	moving vehicles	high-visibility clothing
	chain saws	special protective clothing
	high temperatures	heat-resistant clothing
	inclement weather	all-weather clothing

The following publications give advice on PPE:

- HSE's booklet L25 'Personal Protective Equipment at Work'
- HSE's booklet HS(G) 53 'Respiratory Protective Equipment: A Practical Guide For Users'
- HSE's booklet L101, 'Safe working in confined spaces'.

10.3 Safe use of display screen equipment

The increasing use at work of computers, word processors and other electronic graphic display equipment has brought in its wake a range of health hazards. Requirements to mitigate the effects of these hazards are contained in the *Health and Safety (Display Screen Equipment) Regulations 1992*. These Regulations are supported by an HSE guidance booklet no: L 26 'Display screen equipment work'.

The Regulations:

r.1 Define

- 'display screen equipment' (DSE) as any alpha-numeric or graphic-display screen regardless of the use or process, i.e. not only word processors but computer-design and computer-controlled machinery
- 'operator' as a self-employed user
- 'user' as any person who uses DSE for a significant part of their normal work
- 'work station' as any equipment, or part of it, used in connection with work on a DSE.

r.2 Requires the carrying out of a risk assessment.

r.3 Requires the taking of precautions which are enlarged on in the guidance booklet. The main precautions are summarized below.

Hazards	Precautions
Postural	- adjustable chair so that the operators arms are horizontal when using the keyboard - chair should have adjustable backrest - footrest at a suitable height when the chair has been properly adjusted - adjustable screen so that the operator can look at it in a natural and relaxed head position - document-holder so documents can be held in a position that requires no, or minimal, movement of the operator's head.

r.4 Time
- limit time on continuous keyboard work and ensure breaks at regular intervals. Time between breaks determined by the type of work but should not exceed one hour.

r.5	Visual	• check that the operator has normal eyesight. Indications that this is not so are eye strain, headaches, blurring of the vision
• if not, eyesight tests to be carried out by competent person and the provision of special vision-corrected glasses (employer bears the cost of standard frames but employee contributes extra if they insist on special frames)
• background lighting should be general without points of light that could cause glare on the screen
• the screen should be positioned or adjusted to eliminate glare or reflections from lights or windows
• operators should be able to adjust screen contrast to suit themselves
• screen should be kept clean and free from dust build-up.
r.6	Fatigue/stress	• training in the use of software
• software is suitable for the tasks to be undertaken. Badly designed or inappropriate software can give rise to stress
• effective assistance available quickly when problems met in the use of the programme.
r.7	Information	• hazards associated with the workstation
• precautions provided to avoid hazards.

In addition consideration should also be given to:

Equipment:
- screen should give a stable picture. If instability, flicker, jump, jitter, etc. persist seek advice from supplier
- polarity should be adjustable to suit the user
  positive polarity = dark characters on light ground
  negative polarity = light characters on dark ground
- keyboards should suit the user or the user be trained to the particular keyboard provided
- work desks need to have adequate surface space to accommodate the keyboard, working materials, document-holder, etc.
- surface of work desks should be plain, smooth and easy to clean
- the workstation should be spacious with clear space beneath to allow freedom of leg and body movement.

Environment
- noise should be kept to a minimum. Adjacent noisy equipment (printers, etc.) should be enclosed in soundproof cabinet

- sound-absorbing screens help reduce noise from other equipment
- electronic equipment can make the atmosphere dry; ventilation or other means (plants) should be used to maintain humidity
- radiations – electromagnetic, ionizing and radio frequency – emitted by DSE are well below national recommended levels and less than every-day background levels
- static electricity build-up can cause skin irritation in some people. Reducing the build-up on the screen by wiping it with a damp cloth can help.

User interface
- users must be trained in the particular software
- software must be suitable for the task
- software must be capable of being used at a rate to suit the user
- there should be effective back-up to resolve problems that the user cannot deal with.

Part 4 Safety technology

The need for legislation to give work people protection in the jobs they were forced by circumstances to undertake became a matter of increasing urgency as the Industrial Revolution got under way bringing in its train an enormous toll of horrendous injuries. Machines were being developed to increase the output per worker that paid scant regard to the worker's health or safety.

Initially the major area of concern was the dangerous nature of the machinery but as we became more industrialized and the machinery more complex, a vast range of chemicals has been introduced to aid productivity and to improve the finished product – chemicals whose short and long term health effects are often little understood.

In modern industry and commerce many of the methods of protection have a technical background and a considerable degree on expertise is necessary to ensure those methods are effective against the hazards faced.

This part deals with various of those hazards and the techniques needed to give an acceptable level of protection.

11 Chemicals

Of the vast range of chemicals and substances in use in industry, commerce and other areas of employment, many are harmless. However, there are also a considerable number of substances in use which have a high potential to cause harm and ill-health – 'substances hazardous to health' usually referred to as *hazardous substances*.

There is a considerable range of legislation dealing with chemicals which has as its objective the protection of those who handle these substances by ensuring that employers who make or use these dangerous substances have in place precautions and systems of work that reduce to a minimum the chances of the substances harming anybody.

Excluded are the substances that can cause physical damage, such as by explosions, i.e. flammable gases and explosives, which have separate laws.

It is convenient to consider the problems of the safe use and handling of hazardous substances in three parts which conveniently coincide with the relevant laws:

1 Safe use of substances
   – *The Control of Substances Hazardous to Health Regulations 1994* (COSHH) – see Section 11.1
2 Supply of substances (covering packaging and labelling)
   – *The Chemicals (Hazard Information and Packaging for Supply) Regulations 1994* (CHIP 2) – see Section 11.2
3 Transport of substances by road and rail
   – *The Carriage of Dangerous Goods (Classification, Packaging and Labelling) and Use of Transportable Pressure Receptacles Regulations 1996* – see Section 11.3

Where specific problems arise in the use, handling or transporting of hazardous substances reference should be made to the particular relevant Regulations. Copies of the Regulations and supporting advisory Codes of Practice and Guidance should be held by the local reference library. However if they are not available there, copies can be purchased from HMSO and/or HSE Books.

11.1 Safe use of chemicals

The main legislation concerned with safety in the use of chemicals is *The Control of Substances Hazardous to Health Regulations 1994* (COSHH) which is aimed at reducing the likelihood of ill-health that can result from lax ways of handling and using some of the nastier chemical substances. The main requirements of these Regulations are summarized below.

r.2. Defines:

- a substance hazardous to health as:
  - a substance listed in part 1 of the Approved Supply List – see Section 11.5
  - a substance given an MEL or OES – see Section 11.6
  - a biological agent
  - any dust in substantial concentrations (roughly more than 10 mg/cu m of air which is a thick cloud)
  - any other substance causing a similar hazard to health
- a biological agent as:
  - any micro-organism, cell culture or human endoparasite that may create a risk to human health
- a carcinogen as:
  - any substance likely to cause cancer and classified in the Approved Supply List as carcinogenic
- the maximum exposure limit (MEL) as:
  - the air-based concentration that must not be exceeded – see Section 11.6
- the occupational exposure standard (OES) as:
  - the air-based concentration for other slightly less hazardous substances which the employer should aim to keep as far below as possible – see Section 11.6.

r.3 Places duties on the employer to ensure that chemicals are used and handled in such a way that no one, whether an employee or not, is put at risk.

r.4 Bans:

- the importation (except from an EU country) of:
  - 2-naphthylamine
  - benzidine
  - 4-aminodiphenyl
  - 4-nitrodiphenyl
  - matches made with white phosphorus

- the use in certain processes of:
  - the above substances
  - substances with free silica
  - carbon disulphide
  - mineral oils for use on the spindles of self-acting mules
  - ground or powdered flint or quartz
  - white phosphorus
  - hydrogen cyanide
  - benzene and substances containing it.

r.5 Excludes:

- lead which is subject to the *Control of Lead at Work Regulations 1980*
- asbestos subject to the *Control of Asbestos at Work Regulations 1987*
- substances whose sole risk is physical, i.e. radioactive, explosive or flammable

- substances used for medical treatment
- substances used in mines under mines legislation.

r.6 Requires employers:

- to carry out a risk assessment wherever employees may be subject to risk to health from hazardous substances – see Section 3.4.

r.7 Requires employers to prevent or control the exposure of employees to:

- hazardous substances, except carcinogens or biological agents, by means other than personal protective equipment. [Where biological agents are handled, the only safe protection may be in the use of personal protection in the form of ventilated suits]
- carcinogens where they cannot be replaced by an alternative substance or process, to take the following precautions:
  a) total enclosure of the process and handling system
  b) the use of plant and equipment that minimizes the possibility of leakage or spillage
  c) limit the quantities in use
  d) limit the number of persons exposed
  e) prohibit eating, smoking and drinking in the area
  f) provide adequate hygiene facilities and keep work area clean
  g) identify areas at risk with suitable warning notices
  h) ensure high standards of safety in storage and handling of substances
- if the above measures do not reduce exposure enough, provide appropriate type of personal protective equipment as well
- leakages of a carcinogen or a biological agent when:
  - only authorized person with full protective equipment allowed in the area to effect repairs
  - any employee or other likely to be affected must be told of incident
- substances for which an MEL is specified where exposure must be reduced as far below the MEL as possible
- a substance for which an OES has been approved where exposure must be as low as possible. If the OES is exceeded, the employer must identify the cause and take corrective action.

Any respiratory protective equipment (RPE) provided must:

- comply with the PPE Regulations (see Section 10.2)
- provide suitable protection against the substances concerned
- be approved by the HSE.

r.8 Employers must ensure that all control measures and equipment provided are properly used.
Employees must use protective equipment properly, return it to its store and report any defects or damage found.

130 Chemicals

. . . regularly monitored . . .

r.9 All equipment and plant provided in compliance must be kept in an efficient state, effective working order and good repair. Engineering controls must be examined and tested as follows:

- local exhaust ventilation (LEV)   – every 14 months
- ventilation systems for:
  - blast cleaning of metal castings   – every month
  - grinding and polishing of metals   – every 6 months
  - dust and fume from non-ferrous castings   – every 6 months
  - jute cloth manufacture   – every month

RPE must be examined and tested at suitable intervals.
A record must be kept of the examinations and tests and held for five years.

r.10 Where employees are exposed to hazardous substances, the atmosphere should be monitored regularly and readings recorded. Monitoring can be carried out by:

- 'grab sampling' using a pump to draw air through a stain indicator tube; the length of the stain gives a measure of the concentration. However:
  - measure is not very accurate
  - stain tubes can only be used once
  - stain tubes can only be used for one substance
  - necessary to keep stocks of tubes for each substance monitored
  - can give false reading with mixtures of substances
- passive or activated carbon collector used with air pump:
  - requires complex chemical analysis to obtain concentration

- electronic meters
  - expensive to purchase
  - give immediate and accurate reading in parts per million (ppm)
  - can be used for a range of substances at any one time
  - can be used continuously
  - may include facility to integrate daily exposure levels
  - contain an automatic alarm if pre-set limit exceeded.

Monitoring for:

- vinyl chloride monomer — continuously
- chromium processes — every fourteen days

Records of monitoring must be kept:

- for personal exposure records of identified employees
  — for forty years
- all other readings such as general area monitoring
  — for five years.

r.11 Health surveillance is to be carried out:

- by employment medical adviser or appointed doctor
- every twelve months where an employee works in specified processes with:
  - vinyl chloride monomer
  - nitro or amino derivatives of phenol and benzene
  - potassiun or sodium chromate or dichromate
  - 1-naphthylamine and its salts
  - dianisidine and its salts
  - dichlorbenzidine and its salts
  - auramine
  - magenta
  - carbon disulphide
  - disulphur dichloride
  - benzene and benzol
  - carbon tetrachloride
  - trichloroethylene
  - pitch
- where exposure occurs to any other hazardous substance which could give rise to an identifiable ill-health condition
- records of health surveillance to be kept for forty years
- if the medical adviser bars an employee from working with a particular substance the employer must ensure this happens
- medical examinations should take place during normal working hours
- employees must be allowed to see their medical records
- medical adviser may inspect the work place if he/she wishes
- appeal against a medical decision must be made to the HSE.

r.12 Employees exposed to hazardous substances are to:

- be given such information, instruction and training that they:
  - know the risks faced
  - know what precautions to take

- have access to the results of monitoring of exposures
- have access to collective information from the health surveillance.

Others employed to do the same work must be given the same information instruction and training.

r.13 With certain exceptions, notice must be given to specified authorities of intention to fumigate using:

- hydrogen cyanide
- ethylene oxide
- phosphine
- methyl bromide.

If fumigation is:

- in harbour area, notice to be given to:
  - harbour authority
  - HSE or local authority inspector and
  - if of a sea-going ship:
    * chief fire officer of area
    * officer in charge of Customs and Excise
  - if of a building:
    * chief fire officer of area
- elsewhere, notice to be given to:
  - local police
  - HSE inspector for the area.

Before fumigation starts, warning notices must be posted round the premises concerned.

... give notice of fumigation ...

Chemicals **133**

> r.16 In proceeding for an alleged infringement, a defence is allowed that to avoid causing the breach:
>
> - all reasonable precautions were taken
> - all due diligence was exercised
> - proof must be provided by the accused.

The nub of these Regulations is the identification of hazards from chemicals so that precautionary procedures can be introduced. In addition to professional knowledge of the characteristics of chemicals which may be obtained from the supplier's safety data sheets, the main vehicle for identifying the hazards is the risk assessment (see Section 3.4).

11.2 Labelling of chemicals for supply and use

The *Chemicals (Hazard Information and Packaging for Supply) Regulations 1994* (CHIP 2) which has been amended slightly by the *Chemicals (Hazard Information and Packaging for Supply) Regulations 1996* (CHIP 96) are concerned with ensuring that hazardous chemicals when packaged for sale, both to employers and to consumers, are properly packaged and adequately labelled. Also that safety data on the chemicals are provided to the users.

The content of these Regulations is summarized below.

> r.2 Defines a number of phrases used in the regulations including:
>
> - *approved classification and labelling guide* as the HSC publication 'Approved Guide to the Classification and Labelling of Substances and Preparations Dangerous for Supply' (2nd Edition)
> - *Approved Supply List* as the list entitled 'Information Approved for the Classification and Labelling of Substances and Preparations Dangerous for Supply' (3rd Edition) approved by the Health and Safety Commission on 24 January 1996 (see Section 11.5)
> - *category of danger... for a substance or preparation dangerous for supply...* as one of the following:
>   – physico-chemical properties
>   * explosive
>   * oxidizing
>   * extremely flammable
>   * very flammable
>   * flammable
>   – health effects
>   * very toxic
>   * toxic
>   * harmful
>   * corrosive
>   * irritant
>   * sensitizing
>   * carcinogenic

*mutagenic
   *toxic for reproduction
   *dangerous for the environment
 - *EC number* as:
   - for substances on the approved list, the number quoted
   - for substances not on the approved list, the number given in the European Inventory of Existing Commercial Chemical Substances (EINECS)
   - for new substances, the number given in the European List of Notified Chemical Substances (ELINCS)
 - *indication of danger* as the symbols used to denote the different types of danger shown with an identifying letter. See Section 11.4
 - *preparation* as a mixture or solution of two or more substances
 - *preparation dangerous for supply* is a preparation covered by one or more of the categories of danger listed above
 - *risk phrase* as a phrase describing the risk posed by a substance and designated by the letter R with a number – the phrases and number are given in the Approved Supply List (see Section 11.5)
 - *safety phrase* as a phrase describing the safety precautions to be taken and designated by the letter S with a number – the phrases and numbers are given in the Approved Supply List (see Section 11.5)
 - *substance dangerous for supply* as a substance listed in the Approved Supply List or given one of the categories of danger shown above.

r.3 Specifies that the regulations apply to all substances or preparations dangerous for supply except, *inter alia*:

 - those covered by other legislation such as:
   - ionizing radiations
   - cosmetic products
   - medicines
   - controlled drugs
 - samples taken for examination
 - munitions
 - fireworks
 - LPG.

r.4 Defines the Approved Supply List – see Section 11.5.

r.5 Suppliers must not supply substances or preparations that are dangerous for supply unless they have been classified in accordance with these or other regulations.

r.6 Suppliers of substances or preparations dangerous for supply must provide customers, who intend using the substance at work with safety data sheets containing information on the following:

 - name of the company supplying the substance
 - the substance's name
 - its ingredients

- hazard identification
- first-aid measures
- fire-fighting measures
- accidental release measures
- handling and storage
- exposure controls and personal protection
- physical and chemical properties
- stability and reactivity
- toxicological effects
- ecological effects
- disposal procedures
- transportation requirements
- regulatory matters
- other relevant matters.

Safety data sheets must be kept up to date and given to customers before the first delivery and be in the language of the country of the recipient.

r.7 Advertisements of dangerous substances must mention the danger or hazard.

r.8 Packaging for dangerous substances must:

- be suitable for the substance contained in it
- prevent leakage or escape
- be capable of withstanding expected handling
- have re-usable closures that seal repeatedly.

r.9 Labels must carry the following information:

- name, address and telephone number of supplier
- name of the substance or trade name of preparation
- indication of danger and appropriate hazard symbol
- risk phrases set out in full
- safety phrases set out in full
- EEC number if any
- substance classified as:
  - carcinogenic
  - mutagenic
  - toxic for reproduction

  } must be labelled as *restricted to professional users.*

r.10 Gives information required on labels with specific particulars for:

- pesticides
- paints and varnishes containing lead
- cyanoacrylate based adhesives
- preparations containing:
  - isocyanates
  - epoxy constituents
  - active chlorine
  - cadmium and alloys for soldering or brazing.

Hazardous substances
... properly labelled

> r.11 Sets down details of the printing and attaching of labels.
>
> r.12 Requires:
>
> - child-resistant fastenings to comply with the appropriate harmonized or international standard
> - packages containing a dangerous substance not be of a shape to attract children
> - all packages containing dangerous substances to carry a tactile warning to BS 7280 or EN 272 for the benefit of the partial sighted and the blind.
>
> r.13 Requires a record of the information used to give a substance a classification to be kept for three years.
>
> r.14 Requires notification to the Poisons Advisory Centre of details of any substance classified under 'health effects' above together with copies of the information provided in the appropriate safety data sheet.
>
> r.15 Allows a right of civil action for a breach of these regulations.
>
> Also allows a defence of *took all reasonable precautions and exercised all due diligence to avoid the commission of that offence.*

The above is a very brief summary of the contents of these complex regulations and can form a general guide only. Where a specific problem arises, reference should be made to the regulations themselves.

11.3 Transport of chemicals by road and rail

Whenever dangerous substances are transported on the highway or by rail any incident resulting in spillage of the substance could put, not only the members of the emergency service who have to deal with it, but also members of the public at risk. This risk cannot be entirely eliminated so should an incident occur, information about the substance involved must be immediately available so the emergency services can take the correct action and hence reduce to a minimum the likely effects.

The conditions to be met when carrying dangerous substance by road or rail are contained in *The Carriage of Dangerous Goods (Classification, Packaging and Labelling) and Use of Transportable Pressure Receptacles Regulations 1996* which is in three parts:

1. Introduction
2. Classification, packaging and labelling
3. Transportable gas receptacles

Its contents are summarized below.

Part 1: Introduction

r.2 Defines, *inter alia*:

- a container as being:
  - of more than 1 cu m in volume
  - permanent and capable of re-use
  - designed to facilitate the carriage of goods without intermediate loading
  - readily handleable
  - easy to fill and empty
- dangerous goods as being:
  - explosives
  - radioactive material
  - goods named in the Approved Carriage List
  - any goods having one or more hazardous properties
  - articles or substances.

r.3 Regulations apply to all dangerous goods except:

- fuel in the fuel tanks of vehicles
- goods moved on the orders of the emergency services for repacking or disposal
- explosives
- live animals
- radioactive materials with certain exceptions
- plus a further twelve defined exceptions.

r.4 Lists the information to be contained in the Approved Carriage List (see Section 11.5).

Part 2: Classification, packaging and labelling

r.5 All goods whether dangerous in themselves or containing dangerous components, must be classified in accordance with the Approved Carriage List.

r.6 Where consigned goods are in packages, the package must:

- be designed, constructed, maintained, filled and closed to prevent spillage
- be resistant to the contents
- if fitted with replaceable closure, reclose effectively
- comply with any special conditions imposed by the Approved Carriage List
- be of an approved type, either ADR, RID, UN or joint ADR/RID.

r.7 No package to be marked:

- so as to be confused with an ADR, RID, UN or joint ADR/RID mark
- with an ADR, RID, UN or joint ADR/RID mark unless authorized.

r.8 Particulars to be shown on packages containing dangerous goods are:

- designation of goods (proper shipping name)
- UN number
- the danger sign (see below)
- subsidiary hazard signs.

Packages of mixed goods to carry the label 'Dangerous Goods in Limited Quantities of Classes X, Y, etc.' where X, Y, etc. are the classification codes for the substances in the package.

Retail packages of mixed goods for private domestic use need not display the classification code or UN number but should carry danger and hazard signs.

r.9 + 10 Deal with derogation from labelling requirements.

r.11 Markings and labels on packages should:

- be easily readable
- stand out from the background
- be clearly and indelibly printed
- be securely fixed to the package
- be in English or language of recipient state in EU
- have danger and hazard sign at least 100 mm (4 ins) long.

Part 3: Transportable pressure receptacles (high pressure gas cylinders)

r.12 Pressure receptacles must be:

- manufactured by persons with appropriate technical knowledge
- safe and suitable for use
- comply with approval requirements and relevant EN standards
- not be used if damaged or repaired unless certified safe.

r.13 No transportable pressure receptacle may be used unless it has:

- a written certificate or been stamped:
  - by an approved person or
  - under an approved quality assurance scheme
- an EU verification certificate
- the appropriate marks and inscriptions required the Pressure Vessels Framework Directive.

r.14 Places responsibility on owner to ensure receptacle is properly marked.

r.15 Describes 'approved persons' and requires them to carry out their duties properly.

r.16 Employer must ensure:

- before a receptacle is filled:
  - it has the approved markings
  - it is suitable for the gas with which it is to be filled
  - all necessary safety checks have been made
  - the agreed filling procedure is followed
- after filling:
  - it is within its safe operating limits
  - it is not over-filled
  - any over-fill is removed safely.

r.17 Records to be kept by:

- supplier
- owner of hired out receptacle

of:

- design standard or specification
- EU verification certificate (if relevant).

140 Chemicals

11.3.1
Danger signs

Under r.9, packaging containing dangerous substances must display the appropriate danger signs shown below when being transported.

Compressed gas   Toxic gas   Flammable gas

Flammable liquid   Flammable solid   Spontaneously combustible

Dangerous when wet   Oxidizing agent   Organic peroxide

Infectious substance   Corrosive   Miscellaneous dangerous goods

11.4 Classification of hazardous and dangerous substances for supply

Schedule 1 of CHIP 2 provides a list of the types of danger from chemicals met in the workplace. To each type of danger is given a symbol letter that can be used on labels and warning signs.

Category of danger	Symbol letter
Physio-chemical properties	
Explosive	E
Oxidizing	O
Extremely flammable	F+
Highly flammable	F
Flammable	none
Health effects	
Very toxic	T+
Toxic	T
Harmful	Xn
Corrosive	C
Irritant	Xi
Sensitizing by inhalation	Xn
Sensitizing by skin contact	Xi
Carcinogenic: categories 1 and 2	T
category 3	Xn
Mutagenic: categories 1 and 2	T
category 3	Xn
Toxic for reproduction: categories 1 and 2	T
category 3	Xn
Dangerous for the environment	N

In addition, each of the symbol letters has an equivalent warning symbol

Danger	Letter	Symbol
Explosive	E	
Oxidizing	O	
Extremely flammable	F+	
Highly flammable	F	
Very toxic	T+	

142 Chemicals

Toxic	T	☠
Harmful	Xn	✗
Corrosive	C	
Irritant	Xi	✗
Dangerous for the environment	N	🌳

There are no symbols for dangers from substances that are carcinogenic, mutagenic or toxic for reproduction.

Pesticides are not given separate designatory letters or symbols but must be classified as either very toxic, toxic or harmful, as determined by tests outlined in the Regulations.

The designatory letter and symbol must be displayed on the labels of the packaging of substances dangerous for supply.

11.5 Approved lists	There are two Approved Lists: For supply – *Approved Supply List* – 'Information approved for the classification and labelling of substances and preparations dangerous for supply' (3rd edition) approved by the Health and Safety Commission on 24 January 1996. HSE publication L76. For transport – *Approved Carriage List* – 'Information approved for the carriage of dangerous goods by road and rail other than explosives and radioactive materials'. HSE publication L90.
11.5.1 Approved Supply List	It is divided into seven parts: Part 1 – lists in alphabetical order all those known chemical substances that are considered a danger to health and quotes their index number and CAS number. Non-listing of a substance does not necessarily mean that it is not dangerous. Part 2 – lists mixtures of substances by index number. Part 3 – lists the *R number* (risk number) and corresponding *risk phrase* (see p. 144). Part 4 – lists the *S number* (safety number) and corresponding *safety phrase* (see p. 147).

Part 5 – lists by index number the dangerous substances, giving for each:

- hazard classification details
- specific concentration limits where applicable
- differing risk numbers and phrases where they vary according to the concentration of the substance.

Part 6 – gives information on the classification of pesticides with conventional values of oral LD_{50} mg/kg.

Part 7 – gives additional information for complex coal- and oil-derived substances, listing by index number and quoting CAS number.

The Approved Supply List requires that the information on a label includes:

- hazard symbol
- risk phrase number
- safety phrase number
- EC number
- full name, address and telephone number of the supplier.

11.5.2 Approved Carriage List

This list contains information approved for the carriage of dangerous goods by road and rail other than explosives and radioactive material. It is in two parts, the first listing the substances in alphabetical order and the second listing them in UN number order. In both cases the classification information given to a substance is the same:

Column	Classification information
1	Proper shipping name
2	United Nations number
3	Classification code
4	Subsidiary hazard number
5	Emergency action code for substances carried in tanks.

This is in three parts:

(i) the number indicates suitable fire fighting means:
   1 – water jets
   2 – water fog or fine spray
   3 – foam
   4 – dry agent, i.e. water must not be allowed to come into contact with the substance
(ii) the first letter indicates precautions to be taken
(iii) letter E indicates that local people should be evacuated

6	Hazard identification number
7	Indication of permission to carry in tanks
8	Indication of permission to carry in bulk
9	Packaging group number indicating degree of danger:

   I – great danger
   II – medium danger
   III – minor danger

10	Indication that approval given subject to special conditions.

This classification information should be included on the hazard warning board carried on the vehicle and in the TREMCARD notes carried by the driver.

Risk phrases

Indication of particular risks

R1:	Explosive when dry
2:	Risk of explosion by shock, friction, fire or other sources of ignition
3:	Extreme risk of explosion by shock, friction, fire or other sources of ignition
4:	Forms very sensitive explosive metallic compounds
5:	Heating may cause an explosion
6:	Explosive with or without contact with air
7:	May cause fire
8:	Contact with combustible material may cause fire
9:	Explosive when mixed with combustible material
10:	Flammable
11:	Highly flammable
12:	Extremely flammable
14:	Reacts violently with water
15:	Contact with water liberates extremely flammable gases
16:	Explosive when mixed with oxidizing substances
17:	Spontaneously flammable in air
18:	In use may form flammable/explosive vapour-air mixture
19:	May form explosive peroxides
20:	Harmful by inhalation
21:	Harmful in contact with skin
22:	Harmful if swallowed
23:	Toxic by inhalation
24:	Toxic in contact with skin
25:	Toxic if swallowed
26:	Very toxic by inhalation
27:	Very toxic in contact with skin
28:	Very toxic if swallowed
29:	Contact with water liberates toxic gas
30:	Can become highly flammable in use
31:	Contact with acids liberates toxic gas
32:	Contact with acids liberates very toxic gas
33:	Danger of cumulative effects
34:	Causes burns
35:	Causes severe burns
36:	Irritating to the eyes
37:	Irritating to the respiratory system
38:	Irritating to the skin
39:	Danger of very serious irreversible effects
40:	Possible risk of irreversible effects
41:	Risk of serious damage to eyes
42:	May cause sensitization by inhalation
43:	May cause sensitization by skin contact
44:	Risk of explosion if heated under confinement
45:	May cause cancer
46:	May cause heritable genetic damage
48:	Danger of serious damage to health by prolonged exposure
49:	May cause cancer by inhalation
50:	Very toxic to aquatic organisms
51:	Toxic to aquatic organisms
52:	Harmful to aquatic organisms
53:	May cause long term adverse effects in the aquatic environment

54: Toxic to flora
55: Toxic to fauna
56: Toxic to soil organisms
57: Toxic to bees
58: May cause long term adverse effects in the environment
59: Dangerous for the ozone layer
60: May impair fertility
61: May cause harm to the unborn child
62: Possible risk of impaired fertility
63: Possible risk or harm to the unborn child
64: May cause harm to breastfed babies

Combination of particular risks

14/15: Reacts violently with water, liberating extremely flammable gases
15/29: Contact with water liberates toxic, extremely flammable gas
20/21: Harmful by inhalation and in contact with skin
20/21/22: Harmful by inhalation, in contact with skin and if swallowed
20/22: Harmful by inhalation and if swallowed
21/22: Harmful in contact with skin and if swallowed
23/24: Toxic by inhalation and in contact with skin
23/24/25: Toxic by inhalation, in contact with skin, and if swallowed
23/25: Toxic by inhalation and if swallowed
24/25: Toxic in contact with skin and if swallowed
26/27: Very toxic by inhalation and in contact with skin
26/27/28: Very toxic by inhalation, in contact with skin and if swallowed
26/28: Very toxic by inhalation and if swallowed
27/28: Very toxic in contact with skin and if swallowed
36/37: Irritating to eyes and respiratory system
36/37/38: Irritating to eyes, respiratory system and skin
36/38: Irritating to eyes and skin
37/38: Irritating to respiratory system and skin
39/23: Toxic: danger of very serious irreversible effects through inhalation
39/23/24: Toxic: danger of very serious irreversible effects through inhalation and in contact with skin
39/23/24/25: Toxic: danger of very serious irreversible effects through inhalation, in contact with skin and if swallowed
39/23/25: Toxic: danger of very serious irreversible effects through inhalation and if swallowed
39/24: Toxic: danger of very serious irreversible effects in contact with skin
39/24/25: Toxic: danger of very serious irreversible effects in contact with skin and if swallowed
39/25: Toxic: danger of very serious irreversible effects if swallowed
39/26: Very Toxic: danger of very serious irreversible effects through inhalation
39/26/27: Very Toxic: danger of very serious irreversible effects through inhalation and in contact with skin

39/26/27/28:	Very toxic: danger of very serious irreversible effects through inhalation, in contact with skin and if swallowed
39/26/28:	Very toxic: danger of very serious irreversible effects through inhalation and if swallowed
39/27:	Very toxic: danger of very serious irreversible effects in contact with skin
39/27/28:	Very toxic: danger of very serious irreversible effects in contact with skin and if swallowed
39/28:	Very toxic: danger of very serious irreversible effects if swallowed
40/20:	Harmful: possible risk of irreversible effects through inhalation
40/20/21:	Harmful: possible risk of irreversible effects through inhalation and in contact with skin
40/20/21/22:	Harmful: possible risk of irreversible effects through inhalation, in contact with skin and if swallowed
40/20/22:	Halmful: possible risk of irreversible effects through inhalation and if swallowed
40/22:	Harmful: possible risk of irreversible effects if swallowed
40/21:	Harmful: possible risk of irreversible effects in contact with skin
40/21/22:	Harmful: possible risk of irreversible effects in contact with skin and if swallowed
42/43:	May cause sensitization by inhalation and skin contact
48/20:	Harmful: danger of serious damage to health by prolonged exposure through inhalation
48/20/21:	Harmful: danger of serious damage to health by prolongcd exposure through inhalation and in contact with skin
48/20/21/22:	Harmful: danger of serious damage to health by prolongcd exposure through inhalation, in contact with skin and if swallowed
48/20/22:	Harmful: danger of serious damage to health by prolonged exposure through inhalation and if swallowed
48/21:	Harmful: danger of serious damage to health by prolonged exposure in contact with skin
48/21/22:	Harmful: danger of serious damage to health by prolonged exposure in contact with skin and if swallowed
48/22:	Harmful: danger of serious damage to health by prolonged exposure if swallowed
48/23:	Toxic: danger of serious damage to health by prolonged exposure through inhalation
48/23/24:	Toxic: danger of serious damage to health by prolonged exposure through inhalation and in contact with skin
48/23/24/25:	Toxic: danger of serious damage to health by prolonged exposure through inhalation, in contact with skin and if swallowed
48/23/25:	Toxic: danger of serious damage to health by prolonged exposure through inhalation and if swallowed
48/24:	Toxic: danger of serious damage to health by prolonged exposure on contact with skin
48/24/25:	Toxic: danger of serious damage to health by prolonged exposure in contact with skin and if swallowed
48/25:	Toxic: danger of serious damage to health by prolonged exposure if swallowed

50/53:	Very toxic to aquatic organisms, may cause long-term adverse effects in the aquatic environment
51/53:	Toxic to aquatic organisms, may cause long-term adverse effects in the aquatic environment
52/53:	Harmful to aquatic organisms, may cause long-term adverse effects in the aquatic environment

Safety phrases — Indication of safety precautions

S1:	Keep locked up
2:	Keep out of reach of children
3:	Keep in a cool place
4:	Keep away from living quarters
5:	Keep contents under .... (appropriate liquid to be specified by the manufacturer)
6:	Keep under .... (inert gas to be specified by the manufacturer)
7:	Keep container tightly closed
8:	Keep container dry
9:	Keep container in a well ventilated place
12:	Do not keep the container sealed
13:	Keep away from food, drink and animal feeding stuffs
14:	Keep away from .... (incompatible materials to be indicated by the manufacturer)
15:	Keep away from heat
16:	Keep away from sources of ignition – No smoking
17:	Keep away from combustible material
18:	Handle and open container with care
20:	When using do not eat or drink
21:	When using do not smoke
22:	Do not breathe dust
23:	Do not breathe gas/fumes/vapour/spray (appropriate wording to be specified by the manufacturer)
24:	Avoid contact with the skin
25:	Avoid contact with the eyes
26:	In case of contact with eyes, rinse immediately with plenty of water and seek medical advice
27:	Take off immediately all contaminated clothing
28:	After contact with skin, wash immediately with plenty of .... (to be specified by the manufacturer)
29:	Do not empty into drains
30:	Never add water to this product
33:	Take precautionary measures against static discharges
35:	This material and its container must be disposed of in a safe way
36:	Wear suitable protective clothing
37:	Wear suitable gloves
38:	In case of insufficient ventilation, wear suitable respiratory equipment
39:	Wear eye/face protection
40:	To clean the floor and all objects contaminated by this material use .... (to be specified by the manufacturer)

41:	In case of fire and/or explosion do not breath fumes
42:	During fumigation/spraying wear suitable respiratory equipment (appropriate wording to be specified)
43:	In case of fire, use .... (indicate in the space the precise type of fire fighting equipment. If water increases the risk add – Never use water)
45:	In case of accident or if you feel unwell, seek medical advice immediately (show label where possible)
46:	If swallowed seek medical advice immediately and show this container or label
47:	Keep at temperature not exceeding ....°C (to be specified by the manufacturer)
48:	Keep wetted with .... (appropriate material to be specified by the manufacturer)
49:	Keep only in the original container
50:	Do not mix with .... (to be specified by the manufacturer)
51:	Use only in well ventilated areas
52:	Not recommended for interior use on large surface areas
53:	Avoid exposure – obtain special instruction before use
56:	Dispose of this material and its container to hazardous or special waste collection point
57:	Use appropriate containment to avoid environmental contamination
59:	Refer to manufacturer/supplier for information on recovery/recycling
60:	This material and/or its container must be disposed of as hazardous waste
61:	Avoid release to the environment. Refer to special instructions/safety data sheet
62:	If swallowed, do not induce vomiting: seek medical advice immediately and show this container or label

Combination of safety precautions

1/2:	Keep locked up and out of the reach of children
3/9/14:	Keep in a cool well ventilated place away from .... (incompatible materials to be indicated by manufactuer)
3/9/14/49:	Keep only in the original container in a cool well ventilated place away from .... (incompatible materials to be indicated by the manufacturer)
3/9/49:	Keep only in the original container in a cool well ventilated place
3/14:	Keep in a cool place away from .... (incompatible materials to be indicated by the manufacturer)
3/7:	Keep container tightly closed in a cool place
7/8:	Keep container tightly closed and dry
7/9:	Keep container tightly closed and in a well ventilated place
7/47:	Keep container tightly closed and at a temperature not exeeeding ....°C (to be specified by manufacturer)
20/21:	When using do not eat, drink or smoke
24/25:	Avoid contact with skin and eyes
29/56:	Do not empty into drains, dispose of this material and its container to hazardous or special waste collection point
36/37:	Wear suitable protective clothing and gloves

Chemicals

36/37/39:	Wear suitable protective clothing, gloves and eye/face protection
36/39:	Wear suitable protective clothing and eye/face protection
37/39:	Wear suitable gloves and eye/face protection
47/49:	Keep only in the original container at temperature not exceeding ....°C (to be specified by manufacturer)

11.6 Exposure limits

The limits for acceptable exposure to chemicals hazardous to health through inhalation for the UK are given in the HSE's publication EH40/(latest issue), 'Occupational exposure limits'. This publication is re-issued each year and contains the latest limit values based on the various criteria used.

There are two types of occupational exposure limit which have different criteria:

- *Occupational exposure standards* (OES) for substances that:
  - are unlikely to give rise to a health risk at the exposure levels quoted
  - present no serious long- or short-term health risk within time required to identify and correct any over exposure
  - can reasonably practicably be achieved and improved upon.

- *Maximum exposure limits* (MEL) for substances that:
  - do not satisfy first two of above criteria
  - pose a serious risk to health or
  - meet OES criteria but demand a higher level of control for socio-economic reasons.

In the setting of limits, two groups of specialists are involved;

- Working Group on the Assessment of Toxic Chemicals (WATCH), who:
  - review the chemicals
  - consider toxicological, epidemiological and other data
  - base decisions on scientific judgement of the likely health effects
  - decide if an OES is warranted
  - make recommendations on the level at which an OES should be set
  - if it is decided an MEL is more appropriate, refer the decision to ACTS (see below).

- Advisory Committee on Toxic Substances (ACTS), who:
  - consider:
    * socio-economic implications
    * risk to health
    * cost and effort of reducing exposure
  - set an appropriate MEL level
  - make recommendations to HSC for both OESs and MELs.

Exposure figures quoted are time-based:

- long-term exposure:
  - usually eight-hour period
  - averages exposure over that period (eight-hour time-weighted average [8hr TWA])
  - over-exposure for short periods not likely to give rise to ill effects
  - any over-exposure should not exceed a duration of one hour per day

- short-term exposure:
  - usually fifteen minutes (fifteen-minute reference period)
  - quotes the level at which ill-health effects may result from short-term exposure
  - if no short-term limit quoted, a figure of three times long-term exposure should be used.

Measures of exposure concentrations in air:

- for fumes:
  - parts per million (ppm)
  - milligrammes per cubic metre (mg/cu m)
- for dust:
  - milligrammes per cubic metre (mg/cu m).

Methods for measuring concentrations in air:

- for fumes and vapours:
  - stain detector tubes
    * single reading per tube
    * questionable accuracy
  - passive monitors (carbon discs) ⎫ need complex
  - ventilated carbon filters ⎬ chemical analysis
  - electronic gas monitors:
    * give instantaneous reading
    * continuous reading
    * incorporate warning alarm
    * can give short- and long-term integrated exposure
    * expensive
- for dusts
  - sampling pump with filter or collecting head:
    * needs accurate balance to weigh dust
    * with cellulose dust need to 'condition' filter before and after test.

EH40/(latest issue):

- lists those substances that have been given:
  - MEL
  - OES
- gives methods for calculating:
  - eight-hour time-weighted average exposure for exposure to varying concentrations of the same substance
  - a method for estimating the exposure to mixtures of substances
- gives information on:
  - how the various values have been arrived at
  - an explanations of the hazards presented by particular substances such as:
    * man-made mineral fibres
    * cotton dust
    * asphyxiants
    * lead
    * rubber fumes and rubber-process dust
    * grain dust
    * asbestos
    * carcinogens.

The data given in EH40 is not absolute and should be taken as a guide to the standards to be achieved. It is incumbent upon all employers to endeavour to reduce exposures to below the limits quoted.

11.7 Preventative and control measures

In the use of chemicals, there are a number of measures that can be taken to eliminate the hazard and so prevent employees being put at risk of ill-health. Where the hazard cannot be eliminated control measures must be implemented to reduce to a minimum the risk employees face from the chemicals. The primary aim of these measures should be to protect the whole of the workforce rather than the individual operator who is handling the substance.

It is helpful to follow a strategy when considering control measures:

1 **Identify** the hazards from the substances (see Section 11.4).
2 **Measure** the degree of hazard (concentration of the substance in air – see Section 11.6).
3 **Evaluate** the risk (see Section 3.4).
4 **Implement** preventative and control measures (outlined below).

11.7.1 Preventative measures

Typical preventative measures include:

- Substitution – replacing the particular hazardous substance with one that is non- or less hazardous but which still meets the needs of the process.
- Bulk handling – by handling hazardous chemicals in bulk they can be dispensed automatically through sealed pipelines or by remote conveying. This can have the advantage of greater accuracy of dosing using electronic measuring and control techniques.
- Segregation – by separating the operator from the substances being handled. This can be achieved where manual feeding is necessary by having the substance in pre-weighed sealed bags that can be fed into hoppers that can be sealed. Alternately, where the substance has to be weighed out, this can be done in glove boxes. Another method where the operator needs to move about, is to clothe the operator in a complete ventilated suit with an air supply from a clean source.
- Personal hygiene – much ill-health from chemicals stems from small amounts being carried on clothing or hands and being ingested when eating, drinking or smoking. There should be a ban on eating, drinking or smoking in the work area. Also employees should change their overalls and wash their hands before they eat, drink or smoke.
- Good housekeeping – accumulations of substances on the workplace floor, either as dusts or as residues left over from previous process batches, can contaminate clothing and should be cleaned up for disposal to stores or an appropriate waste bin.
- Eating arrangements – separate accommodation should be provided for the consumption of food and drink and employees should change their overalls and wash their hands before using it. Where facilities for smokers are provided they should be such that they do not interfere with other diners.

11.7.2 Control measures

Where it is not possible to take preventative measures and it is necessary to use a hazardous substance, control measures should be implemented that reduce to a minimum the degree of exposure of the operator to the substance. Typical control measures include:

- Dilution ventilation – where the concentration of a substance in air is around the maximum permitted level (OES) it is possible to reduce that concentration to a safer level by introducing a supply of clean air. Where this is done, checks must be made of the resultant concentrations to ensure the desired reduction is achieved. Dilution ventilation should *not* be used where the substance has been given a Maximum Exposure Level (MEL).
- Local exhaust ventilation – this is a system whereby the hazardous substance in the form of either dust, fume or vapour is extracted at the point of generation. This can be achieved by the use of movable extract hoods placed over the point of generation (e.g. for welding fumes, etc.) or by carrying out the process in an enclosed booth (i.e. spray booth) from which air is extracted. In both cases checks should be made to ensure there is an adequate flow of air. Also periodic examination and testing of the extraction plant is required.
- Reduced time of exposure – in certain circumstances it may be permissible to adjust the operators' work pattern so that the total time of exposure to the substance in a shift ensures that the exposure level, averaged over the shift, is well below the permitted level (OES). This practice should *not* be followed for substances that have been given an MEL.
- Personal protective equipment (PPE) – this should always be considered as a last resort after the above methods have been shown to be ineffective or impracticable. It is essential that the equipment provided is suitable for the substance and the operator and does not interfere with what the operator has to do – see Section 10.2. Practical advice can be obtained from the equipment supplier and from an HSE guidance booklet 'Personal protective equipment at work. Guidance on the Regulations' Booklet no: L25.

11.8 Handling hazardous and dangerous substances

Hazardous substances can be supplied in different ways depending on the quantity involved, from the small retail package to supply in semi-bulk containers and bulk supplies by tanker. Whatever the size of the container, if a dangerous substance or substance hazardous to health is involved, the container must be labelled to identify its contents. (CHIP 2 specifies the sizes of labels for variously sized containers.)

When handling chemicals, whether at the supply stage or during use or disposal, certain procedures should be followed to ensure operator safety. The particular procedure depends on the manner of supply or use and is considered in two parts:

1 Bulk and semi-bulk:

- Delivery:
  - know the safety data for the substance being handled
  - hose connections to have different threads for different substances
  - position warning sign
  - earthing link
  - wear protective clothing

- provide eye-wash facilities
- spillage clearance:
  * by absorbent materials
  * not flush down drain unless approved and then only when well diluted
- follow an agreed system of work
- over-fill:
  * have warnings to prevent
  * have procedure to draw-off surplus
- prepare emergency procedures
- ensure fire precautions are in place
• Use:
  - check condition of pipes for integrity
  - ensure system of work is followed
  - position semi-bulk containers in safe place
  - ensure earth link is connected
  - ensure materials are available to clear a spillage
• Road tankers/transport:
  - should carry hazard warning board at front and rear with:
    * HAZCHEM code
    * substance identification number
    * hazard warning sign
    * telephone number of contact for technical advice
  - driver should:
    * be properly trained
    * be provided with TREMCARD or similar:
        for each substance carried
        giving details of substance
    * ensure TREMCARD is:
        kept in the cab
        readily available
        matches the substance being carried.

2   **Small containers and sacks:**

• Supply:
  - should be palletized
  - check for damage to packages
  - packages should be individually labelled
  - labels to:
    * identify substance
    * indicate hazards
    * give suppliers name and address
• Road transport:
  - to carry orange warning plates at front and rear
• Handling:
  - mechanical if possible
  - manual with barrows or other handling aids
  - manual as a last resort
  - procedure for clearing up spillages
• Use:
  - decant into safe, non-spill type, containers
  - earthing links for both liquids and powders
  - draw quantities for day's use only
  - measures to prevent dust from powders/granules

       - with manual loading:
         * precautions to prevent overcharging causing exothermal or uncontrolled reaction
       - check for explosion risk
       - ensure fire precautions are in place
- Storage:
  - entry to store to have HAZCHEM warning signs
  - ensure adjacent substances are compatible
  - separate incompatible substances, i.e fertilizer and carbonaceous materials; acids and alkalis; etc.
  - ensure every container is properly and clearly labelled
  - record storage positions of each substance
  - provide appropriate fire extinguishers.

Vehicle hazard warning board

Hazard warning signs

Compressed gas   Toxic gas   Flammable gas

Flammable liquid   Flammable solid   Spontaneously combustible

Dangerous when wet

Oxidizing agent

Organic peroxide

Toxic

Corrosive

HAZCHEM sign for mixed storage

11.9 Special waste regulations

Full title is *The Special Waste Regulations 1996* and it is concerned with the safe disposal of wastes that present as particular hazard to health and the ecology.

The Regulations:

> r.2 Define:
>
> - *special waste* as:
>   - those substances listed in Schedule 2 (see below) and
>   - controlled waste containing substances with the following hazardous properties:
>   - highly flammable and flammable liquid
>   - irritant
>   - harmful
>   - toxic and very toxic
>   - carcinogenic
>   - corrosive in concentrations above a certain level and
>   - medical products on prescription only

- Consignor:
  - the person who transfers the waste
- Carrier:
  - the person who transports it from the consignor to the consignee
- Consignee:
  - the person who receives the waste
- Agency:
  - the Environment Agency in England and the Scottish Environment Protection Agency.

r.4 Where waste is to be consigned, the Agency will give that consignment a unique code.

r.5 Outlines the standard consignment procedure based on a five-part consignment note, with five copies to be filled in complete with the unique code.

1. Consignor completes parts A and B of all five copies
2. Consignor sends one copy to the Agency for receiving area
3. Carrier completes part C of remaining four copies
4. Consignor completes part D of remaining four copies and retains one
5. Consignor gives three copies to carrier
6. Carrier gives three copies to consignee on delivery of waste
7. Consignee completes part E of all three copies and:
   - gives one copy to carrier to retain
   - retains one copy
   - sends one copy to the Agency for the receiving area

Parts of the consignment note are:
A – consignment details
B – description of waste
C – carrier's collection certificate
D – consignor's certificate
E – consignee's certificate

r.6 Outlines the procedure to be followed where a carrier collects regular consignments, possibly from a series of consignors (referred to as a 'round')

r.15 Copies of the consignment notes to be kept in registers by:
- consignor on site for three years
- carrier for three years
- consignee on site until his disposal licence terminates when register to be given to the Agency for the area

r.16 A record of the location where each special waste is disposed is to be kept by the consignee until his site licence expires or is terminated, when the records are to be passed to the Agency for the site

r.18 If charged with a breach, it is a defence to plead unable to comply because it was an emergency but that all reasonable steps had been taken to minimize the threat to the public and to comply

Schedule 2 lists special wastes as residues and wastes from various industrial and commercial processes that contain, inter alia:

- agrochemical wastes
- wood preservatives
- degreasing agents
- acid alkyl sludges
- acid tars
- substances containing mercury
- sulphuric, hydrochloric, nitric, hydrofluoric and other acids
- alkaline solutions
- salts and waste containing cyanides, arsenic, mercury, and heavy metals
- asbestos
- inorganic pesticides and biocides
- substances containing halogenated solvents
- organic solvents, washing liquids and mother liquors
- paints, inks, adhesives, varnish, etc containing halogenated solvents
- photographic developer solutions
- slags, dross and flue dusts from aluminium, lead, zinc and copper smelting
- pickling acids
- machining oils and coolants
- oils containing PCBs and PCTs
- brake fluids
- bilge oils
- chlorofluorocarbons and other degreasing solvents
- explosives (ammunition, fireworks, etc.)
- lead and mercury batteries
- sludge from tank cleaning
- human and animal wastes.

Because dangerous chemicals dumped randomly can very quickly leech into a source of drinking water, particular responsibilities are placed on waste site managers through the *Waste Management Licensing Regulations 1994* and on everyone in the waste disposal chain through the *Environmental Protection (Duty of Care) Regulations 1991*.

Other less hazardous waste referred to in EPA as 'controlled waste' is defined in the *Controlled Waste Regulations 1992*.

12 Noise and hearing protection

Hearing is something we take for granted and we subject it to all sorts of excesses. But our ears are very delicate organs and while they will accommodate a certain amount of abuse, they do eventually react and stop working properly. The following sections consider noise legislation, how the ear works and some of the measures that should be taken at work to protect the ability to hear.

When considering noise certain points need to be understood:

- noise is *unwanted* sound
- some sound is essential for:
  - communicating
  - warning
  - balance and orientation
- sounds are pressure pulses in the air
- the threshold of hearing is the lowest level of noise detectable by the ear.

Good hearing is an important feature of a good quality of life – it needs protecting.

12.1 Legislation concerning noise

There are two major pieces of legislation concerning noise, one aimed at protecting employees while they are at work and the other concerned with protecting the quality of life of the community.

12.1.1 Noise at work

The Noise at Work Regulations 1989 outline actions that should be taken to reduce the ill effects of high noise levels at work and puts obligations on employers to take the appropriate action. It incorporates the requirements of the EU Noise Directive.

> r.2 Defines:
> - statutory noise measure as **daily personal noise exposure**, written $L_{EP.d}$, which:
>   - is the integrated noise exposure over the day or shift
>   - can, with difficulty, be calculated from separate noise-level readings
>   - can be obtained accurately using an integrating dose meter called a dosimeter (see Section 12.3)
> - *the first action level* as occurring at a noise level of 85 dB(A) $L_{EP.d}$

- *the second action level* as occurring at a noise level of 90 dB(A) $L_{EP.d}$
- *the peak action level*, which relates to impact noise such as hammering and is a sound pressure of *200 pascals* (equivalent to approx. 140 dB(A)).

Requires employers to:

r.4
- make a *noise assessment* where employees are subjected to excessive noise levels (indicated by having to shout when conversing at arms length)
- ensure noise assessment is carried out by a *competent* person, i.e. someone who has been trained to use a noise meter
- tell employees the results of the assessment
- make further noise assessments whenever changes are made that affect the noise level

r.5
- keep a record of each noise assessment

r.6/7
- reduce noise levels to protect the whole of the workforce rather than provide hearing protection to individuals, i.e. reduce the level of emission at source or contain the source in an acoustic chamber
- reduce noise levels as far as is reasonably practicable

r.8
- make hearing protection (ear muffs or plugs) available on request to employees exposed to the first action level of noise
- issue hearing protection to employees exposed to the second action level of noise

r.9
- in areas where noise levels exceed the second or the peak action levels:
  – designate as *ear protection zones*
  – identify by signs to BS 5378
  – make arrangements known to employees
  – not allow employees to enter unless they are wearing hearing protection (in large machine shops, process plant, etc., where pockets of high noise level occur, identification of individual areas and enforcement of wearing protection may prove problematic and it may be easier to designate the whole shop as an ear protection zone)

r.10
- ensure that all equipment provided to protect employees' hearing is kept in good order and properly maintained

r.11
- give employees information, instruction and training on:
  – levels of noise in the workplace
  – risks to hearing from that noise
  – precautions available to minimize the risk
  – employees' obligations under these Regulations

> r.10 requires all employees to:
>
> - use properly and look after any hearing protection equipment issued to them
> - report any damage or defects to their employer
>
> r.12 requires the manufacturer or supplier of machinery or plant that is likely to create noise levels at or above the first action level to give the purchaser information on the noise emission levels.

12.1.2 Community noise

Noises within a workplace can often interfere with nearby residents' enjoyment of the peace and quiet of their homes. Such noises are referred to as **community noise** and are subject to the *Environmental Protection Act 1990* which:

> s.79(1)(g) Includes as a *statutory nuisance* noise emitted from premises that is prejudicial to health or a nuisance.
>
> s.80(1) Gives powers to local authorities:
>
> - to serve *an abatement notice* which can require the nuisance to be:
>   - abated
>   - prohibited
>   - restricted in its occurrence or recurrence
> - to put time limits for compliance.
>
> s.80(2) Puts the onus for compliance on:
>
> - the creator of the nuisance:
> - the owner or occupier of premises emitting the noise (the employer).
>
> s.80(3) Allows an appeal against an abatement notice within twenty-one days of its being served.
>
> s.80(6) Makes failure to comply an offence subject to a fine up to £20 000.

The measurement of community noise and interpretation of the readings is very complex and is best left to specialists.

For guidance on noise at work, consult the following HSE publications:

- HS(G) 56, Noise at work. noise assessment, information on control. Noise guides nos. 3–8
- HS(G) 109, Control of noise in quarries
- Guidance Note PM56, Noise in pneumatic systems
- Guidance Note EH14, Level of training for technicians making noise surveys

12.2 The ear

The ear is a very delicate organ capable of detecting an enormous range of sounds. Unfortunately it is taken for granted and subjected a great deal of misuse, much of which could be avoided. By understanding how the ear works a better appreciation of the reasons for and techniques of hearing protection can be generated.

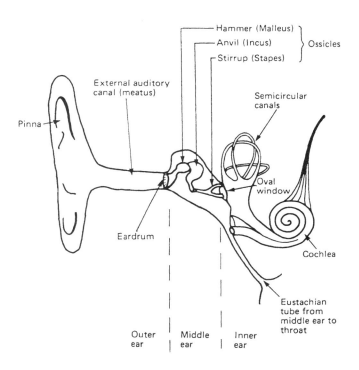

Figure 12.1 Diagram of human ear

Hearing works in the following way:

- Sound pressure pulses enter the outer ear and cause the ear drum to vibrate.
- These vibrations are transferred through the middle ear by a set of three small bones, known as the ossicles, to the oval window of the inner ear.
- The oval window transmits the vibrations to the fluid in the inner ear.
- The fluid carries the vibrations to the sensitive hair cells of the cochlea.
- The hair cells of the cochlea translate the vibrations into electrical signals which are relayed to the brain which then interprets them as sound.

12.2.1 Effects of exposure to excessive noise

Exposure to excessive noise can have the following effects on:

- the ear:
  - permanent damage to the hair cells in the cochlea resulting in:
    * reduced ability to hear (*noise induced hearing loss*)
    * tinnitus (ringing in the ears)
    * a shift in the threshold of hearing with increased difficulty in hearing, particularly noticeable in a crowded room
- behaviour:
  - loss of concentration

- loss of balance and disorientation (due to its effects on the fluid in the semicircular canals)
- fatigue.

While the above relate to occupational noise the same effects can be experienced from leisure noise such as discos and personal hi-fis.

Hearing loss can also result from:

- blockage in the outer ear
- catarrh blocking the eustachian tube causing excess pressure in the middle ear
- a range of medical conditions, some of which may also disturb the sense of balance.

Hearing is a very precious facility that is worth taking trouble to preserve.

12.3 Noise measurement

To know the extent of a noise problem it is necessary to measure the noise levels for which a number of noise level meters of varying accuracy are available and can be used. Similarly, an audiometer can be used to measure a person's hearing ability or acuity.

The **noise meter** is for measuring *noise levels*. There are three basic types:

- general purpose meter
  - relatively cheap
  - sufficiently accurate for identifying areas with noise problems
- grade 1 instrument:
  - gives accurate readings that can be used in noise control measures
  - may include facilities for wave band analysis and integrating exposure levels
  - fairly expensive but necessary if regular noise measurements need to be taken
- precision instrument:
  - measures a range of noise functions
  - gives very accurate readings
  - often linked with recording instruments that measure noise levels over a period of time
  - very expensive and need considerable skill to use.

The **audiometer** is for measuring *hearing acuity*:

- used to measure the threshold of hearing
- indicates hearing loss
- may be manually operated or automatic in taking readings
- records the hearing ability of each ear at a series of different frequencies
- produces an audiogram (a graph of threshold of hearing for each ear at a range of frequencies)*

---

* It is prudent to include an audiogram in pre-employment medical checks.

- test should be taken in an acoustic booth but a quiet room can be satisfactory
- medium cost but only necessary if noise is an on-going problem, otherwise use facility at local hospital.

The **dosimeter** is for measuring daily noise exposure:

- small instrument worn by employee
- consists of small recorder and a microphone which is attached to collar near the ear
- measures and records level of noise every minute of shift
- simple instrument integrates readings to give daily noise exposure $L_{EP.d}$
- more complex instruments allow detailed analysis of recorded data
- analysis requires suitable computer software and data plotter
  - very expensive
  - specialized piece of equipment best left to the specialist
- the only really accurate method of measuring the daily personal noise exposure.

12.3.1 Units of measurement

Noise measurement can be based on either 'sound power level' or 'sound pressure level'. **Sound power level** is the total sound power emitted from a body and is used in measurement of community noise, whereas **sound pressure level** is the level of noise at the point of measurement and is the more common measurement of noise level at work.

The unit of noise measurement is the **decibel**, written dB.

- The decibel is the ratio of measured noise level to minimum detectable noise level.
- It is measured on a logarithmic scale.
- The ear does not interpret noise scientifically but varies according to the frequency.
- Instruments for measuring occupational noise have modified measuring scales ('A' weighting) to match the ear's hearing characteristics, hence the occupational noise unit is dB(A).
- Other weightings exist for particular applications.

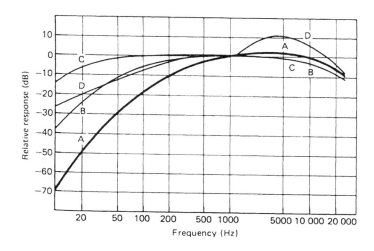

Figure 12.2 Weighting curves

The pitch of a sound depends upon its frequency, i.e. the number of pressure pulses per second, and is recorded as hertz (Hz) or cycles per second. Measurement of frequency is usually based on an octave band and recorded as 'octave band centre frequency'. In an octave band the upper frequency is twice the lower frequency. The octave band centre frequency is 1.414 × lower frequency and typical values are (in Hz):

31.5; 63; 125; 250; 500; 1000; 2000; 4000; 8000

The existence of a noise problem can be identified using a relatively simple noise meter. However, determining the measures necessary to protect against that noise may require more sophisticated and complex equipment requiring specialist knowledge to operate it and interpret the results.

12.4 Noise control techniques

In all hearing protection activites, the first consideration must be to eliminate the source of noise and hence protect all the workforce. However, it is not always possible to eliminate noise emissions completely so action must be taken to reduce emissions as far as possible. This should be done logically following a thought-out strategy.

Two approaches to a hearing protection strategy (each with a listing in order of priority) are:

1   a principles-led approach:

- elimination
  - find alternate method

- isolation
  - remove employee to less noisy area

- insulation
  - enclose noise in sound-insulated container
  - put employee in acoustic cabin

A pragmatic approach to hearing protection

- absorption
  - line walls and reflective surfaces with sound absorbent material
  - use free-standing absorbent panels
  - hang absorbent panels from ceiling/roof
- damping
  - brace or line flat-sheet metal panels to prevent drumming
  - use vibration mounts for machinery
  - use flexible connectors in pipes and ducts
  - use plastic components in machinery
- silencing
  - use silencers on exhaust from air cylinders and vacuum pumps
  - use baffles on outlets from ventilation and extraction systems
  - direct ventilation outlets away from work areas and neighbouring houses (environmental noise)

2 a pragmatic approach:

- engineer out
  - by replacement of plant
  - redesign and modification of plant
  - by altered layout of plant so areas where employees work is at acceptable noise level

- reduce noise at source
  - use of non-metallic components, i.e. plastic gear wheels, rubber bushes in linkages, etc. wherever possible
  - brace or indent metal sheets to stop drumming effect
  - use exhaust silencers, especially on exhausts from air cylinders and vacuum pumps
  - eliminate sharp bends in air and hydraulic systems to stop turbulence noise
  - eliminate electricity frequency hum in transformers – should be installed outside work area but hum can be persistent and penetrating
  - keep working parts in good order by planned maintenance
  - use fans within the makers recommended rating to prevent air drumming
  - etc.

- contain the noise source
  - within an acoustic enclosure
  - must completely enclose noise source
  - need suitable ventilation or plant can overheat and fail

- contain the employees
  - within acoustic cabin or sound haven
  - requires suitable heating and ventilation
  - needs window or other means to view and control process
  - ban Walkmans or other radios

- absorb the noise
  - by use of noise absorbent materials as:
    * linings to walls
    * free-standing panels in work area
    * suspended curtains or panels.

Specialized knowledge of the material to be used and its most effective method of application is required to ensure that noise suppression and control techniques are effective. It is advisable to call in specialists.

A last resort, after all other techniques have proved ineffective, is to provide personal hearing protection which:

- must be personal issue to individual employees
- must provide sufficient attenuation (reduction in amount of noise reaching the ear) to ensure protection of hearing
- users need to get used to different level of sound that can be heard through the hearing protector.

There are two main types:

1. ear muffs:
   - completely enclose each ear
   - must be good seal against the head
   - head band may interfere with other protective equipment
   - special design for use with hard hats
   - types with radio receiver in earcup are suspect since radio noise may interfere with hearing warning sounds
   - can make the ears hot and be uncomfortable

2. ear plugs:
   - permanent type:
     * must be fitted to individual
     * must be kept scrupulously clean or can introduce dirt in ear canal and cause inflammation preventing further use of device
   - throw-away type
     * usually kept in dispensers for employees to take as required
     * one use only
     * cheap but effective
     * some require rolling to shape between thumb and fore finger for insertion in ear – hands must be clean to do this
     * do not interfere with other protective clothing or equipment.

It may be prudent to offer employees a choice of type of personal hearing protection, subject always to the chosen type providing adequate level of protection, i.e. a suitable level of attenuation and being compatible with other protective equipment that has to be worn.

A technique known as 'noise balancing', using electronic equipment to counteract the pressure pulse peaks of noise, is at an experimental stage but it is likely to be many years before it becomes commercially viable and generally available.

13 Work equipment

With the UK's integration into the European Community the influence of Community-based legislation (Directives) on UK laws has become a major factor in current health and safety legislation. This is particularly true with regard to machinery and work equipment.

In 1989 two Directives – the Machinery Directive and the Work Equipment Directive – were adopted and their content required to be incorporated into member states' domestic laws by 1 January 1993. Both these Directives relate to plant and machinery. The UK laws incorporating these Directives are respectively:

- *The Supply of Machinery (Safety) Regulations 1992* (SMSR) – which relates to the safety standards of new plant and machinery which has been purchased since 1 January 1993. One of its aims is to promote the freer movement of goods between Member States using safety standards as the criteria.
- *The Provision and Use of Work Equipment Regulations 1992* (PUWER) – which is concerned with the use of plant and equipment that had been purchased for use at work before 1 January 1993, i.e. existing plant at that date.

These two Regulations are considered in more detail in Sections 13.1 and 13.2 respectively.

13.1 New machinery

All new machinery purchased since 1 January 1993 must comply with the *Supply of Machinery (Safety) Regulations 1992* (SMSR) which incorporate the contents of the EU Directive on the approximation of the laws of the Member States relating to machinery (Machinery Directive, no: 89/392/EEC as amended by Directive 91/368/EEC). These Directives were aimed at removing barriers to the free movement of machinery between member states and used safeguarding as their criteria.

The Directives were drawn up under the 'new approach to legislative harmonization' whereby the main body of the Directive lays down only broad objectives to be met and qualifies these with annexes specifying particular areas or parts to be given consideration which in turn rely on harmonized (EN) standards to detail the techniques to be used as evidence of conformity. SMSR has followed this pattern.

New machinery purchased by any member state must comply with the requirements of the Machinery Directive as evidenced by the CE mark on the equipment together with supporting documentation (technical file). This applies whether a machine is purchased from a UK manufacturer or imported from another EU member state. For machinery imported from a non-EU Member State, the importer is responsible for ensuring the machine has been manufactured to EU requirements and is accompanied by the appropriate documentation (technical file).

168 Work equipment

The Regulations lay down what:

- a supplier must do before putting a machine on the market
- a purchaser can expect when he/she purchases a machine.

The main requirements of these Regulations are summarized below.

r.2 Machinery is defined as:

- *an assembly of linked components, at least one of which moves ... and which are used for the processing, treating, moving or packaging of a material*
- *an assembly of machines which in order to achieve the same end are arranged and controlled so that they function as an integrated whole ...*
- *interchangeable equipment modifying the function of a machine ...*

r.5 Excluded are equipment covered by other Directives including:

- lifting equipment
- manually operated machines
- medical equipment in direct contact with the patient
- fairground equipment
- steam boilers, tanks and pressure vessels
- nuclear plant
- radioactive sources
- firearms
- storage tanks and pipelines of fuels and dangerous substances
- transport vehicles for use on public roads
- sea-going vessels
- passenger cableways
- agricultural and forestry tractors
- machines for military or police purposes.

These Regulations do not apply to:

r.6
- machinery exported to a non-EU country

r.7
- machinery purchased before 1 January 1993

r.12 Puts obligations on a supplier to:

- design/make a machine that satisfies the relevant essential health and safety requirements (ESRs)
- prepare a *technical file* on the machine
- follow the *conformity assessment procedure* shown diagrammatically in Figure 13.2
- issue:
  – either a *Declaration of Conformity* or
  – a *Declaration of Incorporation*
- affix in an indelible manner the CE mark (Figure 13.1)
- ensure the machinery is safe

Figure 13.1 CE mark

> r.14 Lists high-risk machinery as including:
>
> - circular saws for cutting wood and meat
> - sawing machines with manual-feed or loading
> - hand-fed surface planers
> - hand-fed tenoning and vertical spindle moulding machines
> - certain types of underground machinery
> - tractor power take-off shafts
> - lifts for servicing vehicles
> - portable chain saws
> - manually loaded:
>   - thicknessers for wood
>   - band saws for wood or meat
>   - press brakes
>   - injection or compression moulding machines
>   - household refuse vehicles.
>
> For these machines the maker must:
>
> - prepare a technical file and either:
>   - submit it to an approved body for their retention
>   - submit the technical file to an approved body and request:
>   * verification that it meets harmonized standards
>   * a *certificate of adequacy*
> - or:
>   - submit a technical file plus a sample machine for type testing or state where a sample of the machine might be examined.
>
> r.17 Approved bodies:
>
> - are appointed by the Secretary of State
> - are notified to the EU
> - have their name published in the *Official Journal of the European Communities*.

Documentation involved:

r.20 *Certificate of Adequacy*

- issued by an approved body
  - to certify that the machine meets EN standards
  - is presumed to conform with ESRs.

r.21 *EC Type Examination Certificate*

- issued by an approved body
  - to certify that the machine conforms with the appropriate ESRs.

r.22 *Declaration of Conformity*

- by the manufacturer that the machine meets the ESRs.

r.23 *Declaration of Incorporation*

- issued by the manufacturer
- for parts which:
  - comply with the ESRs
  - are to be incorporated into other machinery
  - cannot function independently
  - are not interchangeable.

r.24 The supplier must retain copies of the above documents and the relevant technical file until ten years after the last example of the machine was supplied.

r.25 CE mark to be:

- fixed by responsible person (who can be the manufacturer – so care needs to be exercised when buying from an unknown supplier that the machinery complies – check the supporting documentation)
- distinct, visible, legible and indelible
- in the form shown in Figure 13.1.

13.1.1 Essential safety requirements

These are listed in a schedule to the Regulations and cover a range of items relevant to the safety of machinery. The headings include:

1 Essential safety requirements
   1.1 General
   1.2 Controls
   1.3 Protection from mechanical hazards
   1.4 Required characteristics of guards and protective devices
   1.5 Protection against other hazards
   1.6 Maintenance
   1.7 Indicators

> 2 Additional requirements for certain categories of machinery
>   2.1 Agri-food machinery
>   2.2 Portable hand-held and/or hand-guided machinery
>   2.3 Machinery for working wood and analogous materials
>
> 3 Additional requirements for mobile machinery
>   3.1 General
>   3.2 Work stations
>   3.3 Controls
>   3.4 Protection against mechanical hazards
>   3.5 Protection against other hazards
>   3.6 Indications
>
> 4 Additional requirements for lifting equipment
>   4.1 General remarks including test coefficients (safety factors)
>   4.2 Special requirements for machinery whose power source is other than manual effort
>   4.3 Marking
>
> 5 Additional requirements for underground machinery
>   5.1 Risks due to lack of stability
>   5.2 Movement
>   5.3 Lighting
>   5.4 Control devices
>   5.5 Stopping
>   5.6 Fire
>   5.7 Emission of dust, gases, etc.
>
> 6 Additional requirements for equipment for lifting or moving persons
>   6.1 General
>   6.2 Hazards to persons outside the car
>   6.3 Hazards to persons in the car
>   6.4 Other hazards
>   6.5 Marking
>   6.6 Instructions for use.

Compliance with the conditions of EU directives is through the appropriate UK laws (in this case SMSR) made as part of the UK government's commitment in Europe. However, companies that export may find that customer countries demand compliance with the word of the EU directive for the particular product regardless of whether there are appropriate UK laws or not. Compliance with the directives then becomes a matter of commercial expediency rather than statutory obligation. Any goods exported to or imported from an EU member state must carry the EC mark otherwise they may be rejected.

13.2 Safe use of work equipment

The current legislation that lays down requirements for the use of any equipment at work is the *Provision and Use of Work Equipment Regulations 1992* (PUWER). These Regulations take over the greater part of the requirements contained in the now largely defunct *Factories Act 1961* (FA) but extend to include every item of plant, equipment or tool used at work. However, the standards demanded by PUWER differ very little from those demanded by FA although there are some additional requirements. The parts

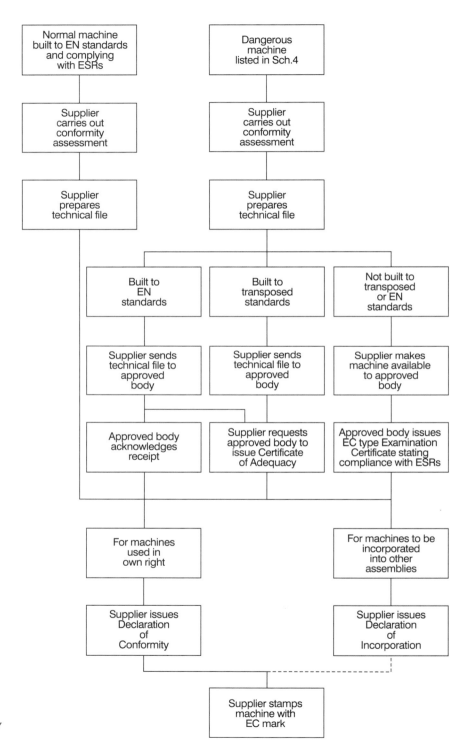

Figure 13.2 Conformity assessment procedure

of FA and its associated Regulations that remain in effect are mainly those dealing with training in the use and the inspection of specific machines.

PUWER applies to the safeguarding and safe use of all equipment, machinery and plant used in any work place. However, where machinery is concerned, any that was purchased after 31 December 1992 should comply with the *Supply of Machinery (Safety) Regulations 1992* (SMSR) (see Section 13.1) which lays down the standard of safeguarding required to enable the machinery to be sold on the open market to EU Member States but is not concerned with safe operating techniques.

The requirements of PUWER are summarized below:

r.2   These Regulations cover anything provided for use at work, from a scalpel to scaffolding, a ruler to a reactor.

r.3   They do not apply to sea-going ships but do apply to offshore gas and oil installations.

r.4   They place obligations on:

- employer
- self-employed
- person who has control of non-domestic premises
- owner or occupier of premises let to individuals for use as a workplace.

r.5   All work equipment to:

- be suitable for the uses to which it is put
- be used only for those purposes.

r.6
- be well maintained (such as to a programme of planned maintenance)
- have a maintenance log.

r.7
- be used only by trained operators
- be maintained only by authorized and trained persons
- have:
  - protective measures to prevent access to rotating stock bars
  - means to stop the machinery if anyone enters a danger area.

r.8   Requires employees to be given:

- adequate health and safety information about the machinery
- written operating instructions about the equipment including
  - limitations of use
  - methods of use
  - foreseeable abnormal situations
  - lessons from earlier use.

r.9   All operators and their supervisors to receive adequate training:

- in the use of the equipment
- risks associated with its use
- precautions to be taken.

174 Work equipment

– provide guards for machines . . .

> r.10 Employers must ensure that new work equipment, i.e. purchased after 1 January 1993, complies with Regulations implementing EU Directives [there is a long list of these Regulations many of which deal with noise, but the main one is SMSR].
>
> r.11 Means must be provided to ensure that:
>
> - access is prevented to dangerous parts of machinery
> - access is prevented to rotating stock bars
> - machinery is stopped before anyone can approach dangerous parts.
>
> Lays down a hierarchy of protective techniques in priority order:
>
> a) fit fixed guards
> b) provide other guards or protective devices such as interlocked guards, trip devices, etc.
> c) use jigs, push sticks or other means that keep the operator and his/her hands clear of dangerous parts
> d) implement a safe system of work with training. [In the UK, systems of work are not acceptable as primary guarding means but only as back-up to other guarding devices.]
>
> Requires that all guards and protective devices:
>
> - are suitable for their intended purpose
> - are of good construction, sound material and adequate strength

- do not create other dangers such as creating traps with moving parts
- cannot easily be by-passed or defeated
- do not allow dangerous parts to be reached
- do not interfere with the operation of the equipment
- permit maintenance to be carried out (see Section 13.4).

r.12 Lists particular hazards where special precautions must be taken to protect employees from:

- articles or substances ejected or falling from machines
- rupture or fracture of parts
- overheating and catching fire
- unexpected discharge of gases, liquids, dusts, etc.
- explosion of machine or product in it
- excludes certain processes covered by other Regulations:
  - lead
  - ionizing radiations
  - asbestos
  - hazardous chemicals
  - noise
  - wearing hard hats.

r.13 Protective clothing to be provided wherever employees are exposed to extremes of temperature, both high and low.

r.14 Requires controls to :

- be clearly labelled to identify the operation controlled.

r.15
- stop and start the machinery, stop control to take priority over other controls
- prevent machine becoming dangerous through over speeding, over pressure, over-heating, etc., whether under manual or automatic control

r.16
- include an emergency stop button
  - must be mushroom-headed, red, with lock-in action requiring positive action to release

r.17
- be operable without putting operator in danger
- be positioned so operator can see if anyone is in a position to be injured if the machine starts. If this is not possible, the controls to incorporate an audible or flashing light warning and delayed-start facility

r.18
- ensure that safety is not prejudiced by:
  - operation of the control system
  - failure in the control system

r.19
- include means for isolating the power supply with locking-off facility.

> r.20 Machines must:
>
> - be securely attached to a base/foundation to make them stable
>
> r.21 - be provided with suitable and sufficient lighting
>
> r.22 - be capable of being maintained safely
>
> r.23 - be marked to show:
>   - safe operating limits, pressure, weight, temperature, etc.
>   - hazards from equipment such as radiation, corrosive substances, heat, cold
> - carry appropriate hazard warning signs [safety sign suppliers will advise on suitable signs].

Existing compliance with the requirements of the Factories Act will give a large measure of compliance with these Regulations.

Guidance on these Regulations is given in:

- 'Work equipment. *Provision and Use of Work Equipment Regulations 1992. Guidance on the Regulations*'. HSE publication number L 22 obtainable from HSE Books.

13.3 Safety with moving machinery

Safety in the use of machinery can conveniently be considered by looking at the hazards that the particular equipment might present. These have been listed in British Standard BS 5304, Safety of Machinery, which, although being overtaken by EU harmonized standards, still gives a good summary of the techniques and practices for ensuring safety in the use of mchinery.

The use of machinery and equipment should be considered for all phases of its life:

- construction
- transport
- installation
- commissioning
- operation including starting up and shutting down
- setting and process change
- cleaning
- adjustment
- maintenance
- de-commissioning and dismantling.

At each phase, one or more of the following hazards may be met:

Hazard	Injury
Entanglement with the machine or material being worked on	shearing or crushing
Contact with the machine or material being worked on	friction burns or abrasions

Hazard	Injury
Being trapped within the machine or between the machine and fixed material or structure	crushing
Being struck by ejected parts of the machine	physical wounds, lacerations
Being struck by ejected work material	physical wounds, lacerations
Contact with sharp edges	cuts
Being drawn in between adjacent parts	crushing
Injection of compressed air or high-pressure hydraulic fluid	emphysema
Contact with electricity	shocks and burns
Contact with hazardous chemicals	ill-health
Contact with hot surfaces	burns
Loud noise	deafness
Prolonged use of vibrating equipment	vibration white finger
Inhalation of mist and fumes	lung disorders, systemic poisoning
Inhalation of dusts	fibrosis, cancer of the lung
Ionizing radiations	skin warts, leukaemia, loss of fertility

Where any of these hazards are identified, action should be taken to eliminate or reduce to a minimum the risk faced by employees. If the hazard cannot be eliminated a risk assessment should be made to determine what safety measures should be taken. Those measures could include the provision of:

- guards
    - criteria:
        * to prevent operators reaching a dangerous part
        * with mesh guards the mesh size will determine what parts of the body can pass through and hence the distance it should be from dangerous parts
    - fixed guards:
        * requiring a tool or key to release it
        * if removable may not be replaced
        * hinged guard bolted or locked shut is more likely to be replaced
    - interlocked guard
        * can be electrical, mechanical, pneumatic or hydraulic
        * must be positive operation, i.e. switch/valve is in relaxed condition when guard is shut (negative operation – actuating spindle of switch/valve is depressed when guard is shut)
        * key-exchange system
        * captive key
        * interlock switches should be of 'fail-safe' design

- adjustable guard
  * with parts that can be adjusted to follow the profile of the work in hand
  * most commonly used on horizontal milling machines
- sleeving
  * the fitting of a loose sleeve round a rotating shaft
- tunnel guard
  * must be long enough to ensure the dangerous part cannot be reached
- automatic guard
  * moves into position automatically when the operation is initiated (i.e. press is struck on)
  * may operate by pushing the operator away from the danger area, sometimes referred to as 'push away guards'
- control guard
  * closure of the guard actuates the machine

- fencing
  - a barrier all round a machine
  - at a distance from danger points greater than operator reach
  - has suitably interlocked access gates
  - for high-risk machines interlock should be key-exchange system

- trips
  - devices which when actuated trip the machine and cause it to stop or revert to a safe condition:
    * push bars
    * pressure mats, edgings or cable
    * photo-electric systems or electro-sensitive protective devices [floor in area covered by beams needs to be marked to prevent inadvertent actuation]
    * pull wires, normally used along the length of a conveyor
    * emergency stop switches
- systems of work
  * not normally recognized as primary means of guarding
  * usually as back-up to guarding systems listed above
  * can only be used if none of the above methods are feasible
  * should be in writing to prevent ambiguity
  * for high-risk work a permit-to-work system should be used.

Guard material can be:

- sheet metal
  - used for fixed guards or where oil splashes/spray needs to be contained
- clear plastic
  - where it is necessary to see behind the guard
    * polycarbonate is tough but soft and easily scratched
    * perspex is harder but more brittle and cracks easily
- mesh
  - allows ventilation cf. of vee-belt drives
    * Expamet needs support frame
    * weld mesh has:
      - greater strength
      - allows better visibility to see beyond guard

- grills
  - can provide considerable physical strength
  - gaps need to be narrow
  - may be visually obstructive.

There are certain necessary operations, such as setting and adjusting, that may need to be carried out with the guards open. This is permitted provided the machine can be moved, either:

- on 'limited inch' only, i.e. actuation of the operating control moves the product 3 ins (75 mm) only then stops and requires a pause before the next actuation

or:

- by 'hold-on' control where:
  - movement occurs only when control actuated
  - release of control stops the machine
  - any movement is at a pre-set crawl speed
  - only one 'hold-on' control is available per machine:
    * either at a number of stations round the machine, only one of which can be selected to be operational at any one time
    * by having single control on a wander lead.

On some special purpose machines such as guillotines with photo-electric safety devices or loose knife cutting machines, additional protection is provided by 'two-hand control':

- the control buttons must be at least 12 ins (300 mm) apart
- both control buttons must be actuated together
- both buttons must be released before the next actuation
- there should be a pause between sequential actuations.

The above describes the main types of guarding that are in general use. However, there are many special purpose guards to suit particular applications which are perfectly acceptable provided they give the necessary degree of protection. Which type of guard is used and how it is applied must depend on the particular machine and the agreed method of operating it.

There is a great deal of guidance available through:

- British Standards
- HSE publications
- industry-based standards
- guard manufacturers.

13.4 Safety during maintenance

Maintenance is one of the more hazardous types of work and particular attention needs to be paid to assessing the risks and preparing for the safe carrying out of the work. The main maintenance areas where high risks arise are:

- work on buildings and roofs at high levels
- power-driven machinery
- in confined spaces such as tanks, vats and underground chambers
- chemical plants.

Common precautions that should be taken whenever maintenance work is contemplated include:

1. Prepare specification for work to be done.
2. Assess potential risks.
3. Prepare work method statement.
4. Prepare detailed plan of work.
5. Develop safe systems of work where necessary including permit-to-work systems.
6. Check the equipment/site to ensure it is safe to start work.
7. Instruct operators and provide training where required.
8. Provide suitable work and safety equipment.
9. Prepare for unforeseen circumstances.
10. Make emergency plans.
11. Monitor that the agreed work methods are being followed.

In addition to the normal hazards of maintenance work, special hazards can arise in particular areas of maintenance work:

Buildings

- working at heights:
  - provide:
    * ladders
    * scaffolding
    * work platforms
    * safety harnesses
- fragile roofs:
  - provide:
    * roof ladders
    * crawler boards
    * edge protection – barrier rails
    * safety harnesses
  - post warning notices
- temporary power supplies:
  - ensure
    * 110-volt supply with centre tapped to earth, or
    * 240-volt protected supply, i.e. with residual current device (RCD)
- access and egress:
  - provide:
    * ladders of adequate length and in good condition
    * walkway boards over scaffolding, floor beams, etc.
    * barriers around all floor openings and floor edges
- hoists and lifts:
  - ensure they have been examined, tested and are in good condition especially in hostile working conditions
  - ensure they are adequate for the loads to be handled
  - where using hired-in lifting equipment, check that:
    * examination and test certificates are current and valid
    * the driver is fully trained and certificated
- inclement weather:
  - provide suitable bad weather clothing
  - warn operators to take extra care
  - provide facilities for drying clothing.

Machinery

- unexpected start-up:
  - isolate power supply by:
    * locking off isolator switch
    * allowing only the person who fits a locking-off padlock to remove it
    * having arrangements to cover shift change-over (a formal hand-over procedure)
  - withdrawing fuses:
    * by electrician who should certify that the machine is safe to work on
    * check by operating the machine starter controls
    * fuses should be kept in secure (locked) cupboard
    * only be replaced by an electrician
- work inside machines:
  - need to certify that the machine is safe to work on
  - where risks are high, implement a permit-to-work procedure
  - arrangements for hand-over of responsibility at shift change
  - if electrical work involved use permit-to-work system
- handling heavy components:
  - estimate weights and ensure adequate lifting tackle is available
  - ensure hoist or cranes have adequate capacity and can give a straight lift
  - if special lifting attachments required, ensure that:
    * they are available
    * the maintenance staff have been trained to use them
  - train staff in any special lifting techniques
  - if hired equipment is used, check:
    * test and examination certificates for validity
    * that the driver is fully trained and certificated
- guards:
  - do not remove until machine certified as safe to work on
  - before allowing machine back into service ensure all guards have been replaced and properly secured.
  - check operation of interlock guards
- explosive/flammable atmospheres:
  - monitor atmosphere
  - use non-ferrous tools
  - if power tools needed use those that are pneumatically-operated
  - use 'inherently safe' or external lighting
  - have fireman standing by with suitable extinguishers
  - prohibit use of naked lights
  - ban smoking
  - prepare an emergency plan
  - consider creating inert atmosphere and providing breathing apparatus
- oxygen-deficient atmosphere:
  - provide:
    * breathing apparatus
    * life-line and safety harness
    * banksman at end of life-line in safe atmosphere.

Confined spaces

- test atmosphere in the confined space using continuous sampling electronic meter with alarm.

- continue testing atmosphere throughout whole time anyone is inside the confined space
  - if no measurable hazard:
    * provide local ventilation from a clean source
  - if atmosphere oxygen-deficient:
    * provide breathing apparatus
    * ensure operators are fully trained in its use
    * ensure safety harness is worn and is attached to life-line
    * station banksman outside confined space to hold life-line
    * ensure the operator can be pulled from confined space in emergency, either by banksmen or using special hoist
    * prepare emergency plan
  - if atmosphere contains toxic fumes or dust:
    * follow procedure as for oxygen-deficient atmosphere (above)
- provide supplies for power and lighting at a safe voltage, 110-volt centre tapped to earth, or 24-volt
- where confined space has inlet pipes for liquids or gases:
  - ensure all inlet valves are shut and locked off [for increased safety on each feed-line, have two valves with interspace vented to drain or atmosphere].

See HSE booklet L101, Approved Code of Practice and Guidance, 'Safe work in confined spaces'.

Chemical plants

Major problems concern leakages from vessels and pipes:

- corrosive liquids:
  - provide 'green' protective suits
  - provide facilities for deluge washing of contaminated suits before operators change clothing
- escaping gases:
  - test atmosphere before starting work

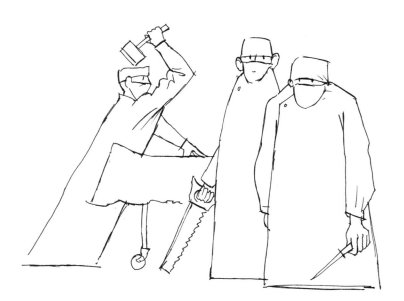

... plan the operation ...

- toxic fumes:
  - provide breathing apparatus
  - since fumes often heavier than air they will flow into depressions and pits, so atmospheres need to be tested before entering the area
- flammable or explosive gases and fumes:
  - monitor atmosphere
  - no naked lights or smoking
  - use non-ferrous tools.

While maintenance work has a bad safety record, the risks to operators can be minimized by planning each operation and ensuring the correct work methods are followed and the necessary safety equipment is provided and used.

13.5 Pressure systems

The legislative requirements for pressure vessels and plant are contained in the *Pressure Systems and Transportable Gas Containers Regulations 1989* with a supporting Approved Code of Practice 'Safety of pressure systems. Pressure Systems and Transportable Gas Container Regulations 1989. Approved Code of Practice'. HSC publication no: COP 37.

These Regulations repeal those parts of the *Factories Act 1961* concerned with the construction and use of steam boilers, steam receivers, steam containers and air receivers but leave unaltered the restrictions on entry into boilers (s. 34). They are pragmatic and recognize that differing operating conditions have differing examination requirements.

The term *pressure system* refers to the installations found in many workplaces for supplying compressed air, steam for heating and processes, hydraulic systems, various liquids, oxygen and gases for welding or burning. It includes all pipework, vessels, heat exchangers, calorifiers, boilers, etc. *Transportable gas containers* refer to the bottles of pressurized gases used for burning and welding operations.

The contents of the Regulations are summarized below.

> r.2 Pressure systems are those that contain a **relevant fluid**, which is defined as:
>
> - steam
> - fluids (single or mixed) at a gauge pressure > 0.5 bar (7.5 psig) such as:
>   - a gas
>   - a liquid with a vapour pressure > 0.5 bar at its stored temperature
>   - a gas that is stored by being dissolved in a porous substance and is released without heating, such as acetylene.
>
> Excluded are pressure systems:
>
> - in ships
> - in weapon systems
> - of a wheeled, tracked or rail-mounted vehicle

- when carried in a vehicle if they comply with international Conventions and Rules
- any pipeline where the gauge pressure < 2 bar
- parts of pressurized workings
- gas containers in motor vehicles or trailers if for propulsion, heating or cooking
- water cooling systems of engines and compressors
- two-part beer kegs
- portable pressurized fire extinguishers
- hand-held pneumatic tools, etc.
- diving equipment.

r.4 Pressure systems must be:

- properly designed and constructed from suitable materials
- capable of being examined safely
- provided with suitable protective devices such as pressure gauges, pressure relief valves, water level gauges, low water alarms, drains, blow-down arrangements, etc.
- safe after being modified or changed.

r.5 Vessels in a pressure system must have a fixed plate with:

- maker's name
- serial number
- date of manufacture
- standard to which built
- maximum design pressure or minimum if a vacuum vessel
- design operating temperature
- plus be marked (by painting) with:
  - the maximum working pressure
  - plant number (this helps identification during examinations and testing).

r.6 The installer of a system is responsible for ensuring it is safe to use.

r.7 A pressure system must not be used until:

- it has been examined
- its safe operating limits are known
- a written scheme of examination has been prepared by a competent person who can be an employee or a consultant. [It is usual for the company that insures the plant to assist in drawing up a scheme – they will probably be carrying out the examination as part of the service under the insurance policy.]

r.8 The scheme of examination should:

- list all protective devices
- consider all parts of the system where failure could cause danger
- state the type of examination, whether visual, internal or test
- outline preparations necessary for the examination

- lay down the frequency of examinations
- be reviewed periodically and amended as appropriate.

r.9 Examinations:

- must be carried out at the times detailed in the scheme
- recorded in a written report:
  - made by the examiner
  - handed to the user or owner
  - which states:
    * parts examined, their condition and results of the examination
    * any repairs or modifications needed and the time limit for completion
    * if condition of system warrants it, a revised safe operating pressure
    * date by which the next examination must be carried out
    * either that the system is safe to continue in present operation or give the reasons if changes are necessary.

r.10 If system found to be dangerous, a copy of the report to be sent to the enforcing authority within fourteen days.

r.11 The user of a pressure system must:

- give adequate instruction to operators to ensure:
  - they know:
    * how to operate the system safely
    * the action to take in an emergency
- check that the system is being used in accordance with instructions

r.12 ensure it is maintained in good repair

r.9
- not use equipment until any necessary repairs have been completed

r.13
- keep copies of:
  - reports of examinations
  - earlier reports that refer to:
    * safe operating parameters
    * any required repairs or modifications
  - information from the supplier regarding examinations
  - reports:
    * on the premises
    * readily available for inspection. They can be on disc but must be:
      - capable of providing a hard copy when required
      - secure from loss or unauthorized interference
      - be authenticated only by the examiner
      - be transferred to any new owner.

Special requirements for *transportable gas containers* include:

r.16
- They are not to be taken into use unless the design standard has been verified and stamped on the container.
- The container must be an EU type with an EU Verification Certificate.

r.17 Before filling a container, the filler must:

- check that it has been properly examined
- ensure it is suitable for the fluid with which it is to be filled
- make any other safety checks
- not over-fill. If he does, he must ensure excess fluid is removed safely
- check after filling that it is within its safe operating limits
- not refill a non-refillable container.

r.18 Containers are to be examined at suitable intervals by a competent person who should mark (stamp) the date of the examination on the container.

r.19 They should not be modified by an employer.

r.20 If modified or repaired, they must not be used until certified by a person or body approved by HSE.

r.21 They can only be re-rated by a person or body approved by HSE.

r.22 Records of:

– design specification or – design standard or – EU Verification Certificate	to be kept by the manufacturer and by the owner of a hired out container
– the tare weight including porous substance and acetone or other solvent of an acetylene container	to be kept by the owner.

The gas contained by these vessels is at very high pressure and damage to a vessel can result in a failure of the pressure containment and explosive discharge of the gas.

13.6 Lifting equipment

The current position regarding the construction and use of lifts is contained in new regulations *The Lift Regulations 1997* incorporating the content of an EC Directive (no: 95/16/EC) *on the approximation of the laws of Member States relating to lifts*. This is an Article 100A Directive concerned with supply and aimed at removing barriers to trade. It is concerned only with the construction of lifts and does not concern itself with the use, maintenance or periodic inspections. Those requirements are contained in sections of the *Factories Act 1961* (FA) that remain extant. However, that too is likely to change since the HSC is considering new regulations concerning the safe operation of lifts.

Under the new regulations referred to above, there is a run-in period to 30 June 1999 during which period lifts can be manufactured to the requirements of either the FA or the new regulations. After that date all new lifts must conform to the new regulations.

Lifting equipment divides into two broad categories: lifts and hoists on the one hand and cranes and lifting tackle on the other.

Lifts and **hoists** are generally described in s.25 of the Factories Act (FA) as:

a platform or cage the direction of movement of which is restricted by a guide or guides

and under the new Lift Regulations as:

an appliance serving specific levels, having a car moving:
(a) along guides that are rigid; or
(b) along a fixed course even where it does not move along guides which are rigid (for example a scissor lift)

and inclined at an angle of more than 15 degrees to the horizontal and intended for the transport of:

- *persons*
- *persons and goods*
- *goods alone if the car is accessible, that is to say, a person may enter it without difficulty, and fitted with controls situated inside the car or within reach of a person inside.*

Lifts for raising goods generally have less stringent safety requirements than those for carrying people.

Cranes are referred to in s.27(9) of FA as being *lifting machines* and include crane, crab, winch, teagle, pulley block, gin wheel, transporter and runway while the *Supply of Machinery (Safety) Regulations 1992* (SMSR) just refer to *lifting machines* without further description.

Lifting tackle is referred to in s.26(9) of the FA as including chain slings, rope slings, rings, hooks, shackles and swivels. SMSR refers to **lifting accessories** as components or equipment not attached to the machine and placed between the machinery and the load or on the load in order to attach it to the hook of the crane, etc. It also refers to **separate lifting accessories** as accessories which help to make up or use a slinging device, such as eyehooks, shackles, rings, eyebolts, etc.

Thus there are four separate types of lifting equipment:

1 lifts for goods
2 lifts for people
3 cranes or lifting machines
4 lifting tackle or lifting accessories.

The safety requirements in respect of all these are dealt with in Sections 16.1 and 16.2.

14 Construction

The construction industry has an unenviable record of accidents including many fatalities. While the type of work carried out does present a high level of hazard, there seems to have been an attitude endemic in the industry that it is all part and parcel of the job. Some recent major construction projects have disproved that contention and have been successfully completed without fatality or serious injury. Much of this success is due to changing attitudes particularly at senior management levels. Whether this is a reaction to new laws or whether the new laws have recognized the importance of the management aspects of construction projects, there is an increasing emphasis in legislation on the wider responsibilities of senior managers and of other off-site people such as architects who play a crucial role in determining on-site safety. These new legislative requirements are embodied in the *Construction (Design and Management) Regulations 1994* (CDM).

Another development in the laws relating to construction has been the rationalization of three parts of the old Construction Regulations into the *Construction (Health, Safety and Welfare) Regulations 1996*.

The contents of both these Regulations are outlined in the following sections which also include advice on a number of common practices within the construction industry.

14.1 CDM

CDM are the initial letters of the *Construction (Design and Management) Regulations 1994*, which place responsibilities for safety on construction sites on everyone involved, from the designer and the client to principal and sub-contractors.

The main requirements of these Regulations are outlined below.

> r.2 The Regulations relate to:
>
> - construction work including *carrying out any building, civil engineering or engineering construction work* where more than five persons are employed
> - all demolition work regardless of how many employed.
>
> r.7 HSE to be notified:
>
> - before any work commences on site
> - if construction work will last more than thirty days
> - if work involves more than 500 person days of construction work

- if more than five persons are employed on site at any one time
- using Form 10(rev) or other form that gives:

    1. Date of sending form.
    2. Exact address of site.
    3. Name and address of client.
    4. Type of project.
    5. Name and address of planning supervisor.
    6. Declaration of appointment by planning supervisor.
    7. Name and address of principal contractor.
    8. Declaration of appointment by principal contractor.
    9. Date for start of construction work.
    10. Duration of construction work.
    11. Maximum number of people on site at any time.
    12. Number of (sub-)contractors on site.
    13. Name and address of (sub-)contractors already chosen.

r.4 Client may appoint an agent to act for him:

- appointment must be:
  - in writing
  - signed by agent
  - copy sent to HSE.

r.5 For domestic developments, the client may appoint the *developer* as agent.

r.6 Client to appoint:

- planning supervisor
- principal contractor

can be client or another person provided appointee is competent for both jobs.

r.8 Client to:

- appoint:
  - planning supervisor
  - principal contractor
  - designer.

r.9 Client to:

- satisfy himself:
  - as to the competence of the above appointees
  - that they have the necessary resources, facilities and experience.

r.10/12 Client to:

- ensure a *health and safety plan* is:
  - prepared before work starts on site
  - kept available for reference.

r.11 Client must:

- give the planning supervisor information about the *state and condition of the premises* where work is to be carried out
- give him adequate resources and facilities to do his job.

r.13 Designer must:

- ensure the client is aware of his (the client's) duties
- ensure that his design will:
  - allow the work to be carried out safely
  - not be hazardous to construct
  - incorporate safeguards for construction workers
  - give information about any hazardous materials or equipment to be used in the work
- co-operate with the planning supervisor.

r.14 Planning supervisor to:

- ensure design of building is adequate
- co-operate with the designer
- ensure a safety plan has been prepared before contractors start work
- be able to give advice to client and contractor to enable them to comply with current legislative requirements
- prepare a health and safety file and deliver or make it available to the client
- be available for consultation by employees of contractor and sub-contractors.

r.15 The safety plan to include:

- description of work involved in the project
- project programme
- information on known or foreseeable hazards
- welfare arrangements
- any other information necessary for contractor to work safely
- information from principal contractor on:
  - site safety arrangements
  - site welfare facilities.

[The safety file can additionally include:

- copy of safety plan
- brief summary of the project
- list of contractors and sub-contractors
- information from main and sub-contractors about, or copies of their:
  - safety policies
  - PL (Public Liability, or third-party) and other relevant insurance cover
  - method statement of how the work will be carried out
  - test and examination certificates of any lifting equipment, whether hired in or owned

– certificates of competence (licences) of crane drivers, truck drivers, scaffolders, etc.
– notes of meetings between client and contractor.]

> r.16 Principal contractor to ensure that:
>
> - (sub-)contractor's
>   – work is co-ordinated
>   – complies with safety rules in the health and safety plan
> - only authorized persons are allowed on site
> - the safety supervisor is given information required for:
>   – the safety file
>   – advising the client of safety matters
>   – ensuring co-operation between the main and sub-contractors
> - the site is secure from unauthorized visitors
> - all (sub-)contractors:
>   – co-operate
>   – comply with safety rules
>
> r.17 – are given sufficient information and training to ensure their health and safety on site
>
> r.18 – are consulted about health and safety matters on site
>
> r.19 (Sub-)contractors are required to:
>
> - comply with safety rules
> - inform the principal contractor of any hazards arising from their particular work.

The main thrust of these Regulations is at the management of construction projects to ensure that the work is properly planned, hazards and risks anticipated, and that adequate and suitable facilities are available to protect the health and safety of anyone working on the site.

14.2 Construction, health, safety and welfare

The *Construction (Health, Safety and Welfare) Regulations 1996* rationalize into one Regulation the contents of:

- *The Construction (General Provisions) Regulations 1961*
- *The Construction (Working Places) Regulations 1966*
- *The Construction (Health and Welfare) Regulations 1966*

which are revoked. Meeting the requirements of these latter Regulations will go a long way towards achieving compliance with the new Regulations which bring to construction work the standards required by the *Workplace (Health, Safety and Welfare) Regulations 1992*.

The duties summarized below deal only with safe working on normal construction sites and only where they refer to matters not covered by the *Workplace (Health, Safety and Welfare) Regulations 1992*. They do not deal with particular duties covered in the *Construction (Design and Management) Regulations 1994,* which are covered in Section 14.1, nor to the precautions necessary in specialized construction work.

r.4 These Regulations place duties on:

- the main contractor
- all sub-contractors
- anyone who controls work on a site
- workers employed on a site.

r.5 They require places of work to:

- be kept safe
- have safe access to and egress from
- be secure from entry by unauthorized persons
- have adequate working spaces
- be well ventilated
- at reasonable temperature
- be adequately lit
- be kept clean and tidy.

r.6 Require provision to prevent falls by:

- use of working platforms
- when liable to fall 2 m or more, provision of:
  - guard rails at least 910 mm (36 ins) from floor
  - intermediate rails to ensure vertical gaps do not exceed 470 mm (18 ins)
  - toe boards 150 mm (6 ins) high
- safety harnesses or safety nets when working at heights.

Ladders must:

- be well maintained
- rest on solid level ground
- if 3 m or more in length be footed and/or tied at the top
- rise at least 1 m above platform served
- if total rise more than 9 m, be provided with intermediate platforms
- have feet placed 1 unit from wall for every 4 units rise.

r.7 For work on fragile roofs:

- access not permitted unless suitable platforms (crawler boards) are provided
- warning notices to be posted.

r.8 Where materials are stacked or stored at a height

- suitable guard rails, toe boards or other means (nets, plastic webbing, etc.) must be provided to prevent it falling
- hard hats must be worn.

r.9 Structures, whether scaffolding, shuttering, timbering or sheeting in excavations must be:

- strong enough not to collapse accidentally
- stable when carrying intended materials
- erected and dismantled under supervision of competent person.

r.10 Demolition must be carried out safely (see Section 14.6).

r.11 Explosives only to be used if no one exposed to risk of injury.

r.12 Excavations:

- steps must be taken to ensure that:
  - excavations do not collapse accidentally
  - no one is liable to be buried or trapped by falling or dislodged material [i.e. by shoring, battering or benching the sides]
- must be supported by suitable equipment:
  - installed, changed and dismantled under supervision of competent person
- vehicles, materials or persons should not approach excavations where they could:
  - fall in
  - cause the edge to collapse
- should avoid or prevent damage to underground services.

r.13 Coffer dams should be:

- of sufficient strength and capacity
- constructed, altered and dismantled under supervision of a competent person.

r.14 Where there is a risk of falling into water, whether pond, lake, river or sea, precautions must be taken to:

- prevent falls
- rescue persons and minimize risk of drowning, this may involve a rescue boat being on permanent patrol
- provide and keep in good order suitable rescue equipment.

Where workers are transported by boat, it should be:

- of suitable size and construction
- properly maintained
- under the control of a competent person
- not overloaded.

r.15 Traffic routes must be:

- suitable for the traffic to be carried
- separate from pedestrian routes
- well signposted
- provided with barriers where pedestrian exits let straight onto traffic route.

r.16 Doors and gates – see Section 5.1.

r.17 Vehicles

- should be parked in a safe manner (handbrake on and engine off)
- should be suitable for the job they have to do
- drivers must:
  - be fully trained and certificated
  - not carry passengers unless the vehicle has suitable seats
  - not man the vehicle during loading/unloading unless the driving position is safe
  - avoid the edges of pits, excavations, earthwork, etc.
- pits, excavations, earthworks, etc. must have means to prevent vehicles overrunning.

r.18 Precautions to be taken to prevent injury from:

- fire or explosion
- flooding
- asphyxiation.

r.19 Emergency escape routes must be:

- kept clear
- known to all working on the site
- clearly identified.

r.20 Emergency plans must be:

- prepared
- made known to all on site
- be practised.

r.21 Site to have:

- suitable fire detectors and alarms
- extinguishers and fire fighting equipment which are:
  - properly maintained
  - clearly identified
- fire training for all workers.

r.22 Welfare facilities to include:

- suitable number of toilets
- washing facilities with hot and cold water, soap, towels, etc.
- supply of wholesome drinking water
- separate accommodation for work and personal clothing
- rest room

r.23
- adequate supply of fresh air

r.24
- reasonable temperature in buildings
- provision of bad weather clothing

r.25 • adequate lighting throughout the site.

r.26 Site to be kept tidy and clean.

r.27 Plant and equipment:

- to be kept in good order
- used only for what it was designed to do.

r.28 Where technical knowledge is necessary for safety, suitable training to be provided.

r.29 Inspections by competent person required of:

- working platforms
- scaffolding
- excavations
- coffer dams and caissons

at following specified intervals:

- working platforms and scaffolding:
  - before first use
  - after changes
  - after damage
  - at least every seven days
- excavations:
  - before each shift
  - after damage or changes
  - after falls of rocks, earth, etc.
- coffer dams:
  - before each shift
  - after damage or changes.

r.30 Reports to be made of all inspections, and kept:

- at the site
- for three months
- available for inspection.

Although these requirements appear extensive, they are little greater than the sum of the three Regulations they replace. Compliance with the earlier Regulations will give compliance with the relevant parts of these replacing Regulations.

14.3 Construction safety

Whenever construction work is to be undertaken if consideration is given to some of the hazards faced and the appropriate precautions taken, accidents and damage can be prevented. This applies whether you are a client having the work done or a contractor doing the work. However, once having agreed precautionary measures it is essential that the work is monitored to ensure the precautions are implemented. The contractor should supply *all* plant and materials needed for the contract.

Typical of the points to be considered are:

- Falls from heights:
  - working platforms:
    * above 2 m from ground/floor, must be provided with hand rail at 1 m, intermediate rail at 0.5 m and toe board 150 mm high
  - for roof work, provide:
    * edge protection
    * safety harness
    * crawler boards if a fragile roof (such as asbestos sheeting)
  - ladders must:
    * be in good condition
    * be inspected regularly
    * be lashed at top end to structure. (If this is not possible post someone at bottom to 'foot' the ladder)
    * if used as access, project at least 1 m above platform served unless safe handhold provided
    * be set no steeper than 75°, i.e. 1 unit out for every 4 units rise
- Falling objects:
  - hard hats must be worn
  - when material stored at high level, ensure platform is:
    * wide enough to allow access passed storage position
    * strong enough to support the weight
    * provided with hand rails, toe board and netting or similar to retain materials

- Trenches and excavations deeper than 1.2 m (4 ft) should:
  - have sides:
    * battered (sloping) or
    * benched (stepped) or
    * properly shored under supervision of competent person
  - be inspected very day
  - be far enough from existing buildings not to affect foundations
  - have edges protected to prevent:
    * people falling in
    * vehicles falling in
  - have ladder for access and egress
  - have arrangements for removing water
- Temporary wiring
  - should be secured to structure, not left hanging in loops
  - connections should be properly made, not taped
  - should be kept for minimum period then removed
  - should be 110-volt, centre tapped to earth (see Section 17.3)
  - if 240-volt, should be protected by residual current device (RCD)
- Power tools
  - should be preferably 110-volt
  - should be regularly inspected, including leads, sockets, etc.
- Materials
  - follow supplier's safety instructions
  - provide PPE (gloves, goggles, face masks, etc.) as appropriate
  - where solvents are used (in adhesives, paint, etc.) ensure work area is well ventilated
- Housekeeping
  - ensure site kept tidy
  - do not allow build-up of rubbish – it is a fire and health hazard
  - rubbish to be removed not burnt on site
  - dust and fumes to be kept to a minimum
- Noisy equipment
  - used only during 'social' hours (7am to 7pm)
  - provide hearing protection to operators and those working nearby
- Overhead power lines
  - indicate by lines of bunting or flags
  - post warning notice
- Underground services
  - check with local authority and gas, water, electricity and telephone companies before excavating
  - check for service runs with suitable instrument
- Scaffolding
  - must be erected by competent certificated (CITB) erectors
  - must be inspected:
    * before use
    * every week
    * after damage or alterations
- Asbestos
  - check type – local analytical chemist can do this
  - get atmosphere monitored to determine concentrations
  - if chrysotile (blue) or amosite (brown) arrange for an approved contractor to strip it out
  - if other types, decide action, i.e. remove, seal or encase
  - use specialist contractor
- Pneumatic tools
  - breakers, chisels, etc.

- risk of vibration white finger (VWF)
- if diagnosed move worker to other work
- check with supplier of tools for availability of attachments to reduce the vibration effect (insulated handles, etc.)
- provide hearing protection for operator and those working nearby
- Mobile cranes
  - ensure test/examination certificates are current
  - ensure driver is properly trained and certificated (CITB)
  - use outriggers when lifting
  - ensure outrigger feet rest on suitable base plates
  - ensure base plates are on solid compacted ground
  - allow room for swing of counterbalance weight
  - use a banksman to assist driver with slinging and lifts
  - if crane hired in:
    * check test/examination certificate of crane and associated equipment
    * check driver is trained and certificated
    * if in doubt refuse to accept the hire
- Welfare facilities
  - each contractor should provide their own but can share by arrangement
  - facilities include toilets, washrooms, canteen, first aid, etc.
- Site arrangements
  - should be made known to all contractors:
    * routes to be taken by workmen and vehicles
    * security arrangements
    * fire precautions and alarm
- Equipment
  - each contractor should provide their own
  - if contractor wants to use another's equipment:
    * agree in writing
    * contractor to give written statement that equipment is in good order
    * if equipment needs driver, either owner provides one or contractor's driver must prove competence (by training certificate)
    * contractor must sign indemnity accepting responsibility for any damage caused by or to equipment during period of hire/loan
- Pits, openings, and platform edges to be provided with a substantial barrier
- Use of local services:
  - agree with site agent before work starts:
    * which services
    * for how long
    * what charge to be made.

Establishing good communications with site agent and between contractors, through the nomination of individuals to be the points of contact, will enable many of the day-to-day problems to be sorted as the work progresses and ensure it is carried out in a much safer manner.

14.4 Employing contractors

At some stage in their existence most companies employ contractors to carry out work on their premises. This work can range from building, electrical installation, installing and commissioning a new plant, etc., through to cutting the grass and cleaning windows.

While the contractor is on the premises the occupier has both statutory and common law responsibilties for his health and safety and that of his employees, but may not have any control over how those employees work and behave. By the same token, the contractor has obligations to ensure that the occupier's employees are not put at risk from the way he carries out the work covered by the contract.

The safe and successful carrying through of a contract depends on good relations and communications between the occupier and the contractor. This is helped by having well defined and understood conditions in the contract.

Occupier's responsibilities include:

- ensuring contract includes conditions requiring the contractor to comply with all relevant statutory health and safety legislation
- informing the contractor of any special hazards on the site
- providing training for the contractor, his employees and sub-contractors in the precautions to be taken
- providing copies of the local safety rules and requiring the contractor, sub-contractor and their employees to comply with them
- identifying the area where the work is to be carried out and over which the contractor will have control
- appointing a planning supervisor to be the point of contact with the contractor
- not allowing factory equipment to be used by the contractor without prior written agreement
- checking that the contractor has adequate Employers Liability (EL) and Public Liability (PL, or third-party) insurance cover. EL cover is mandatory; PL cover should match likely losses (minimum of £1m)
- providing safe access for the contractor's vehicles and employees
- informing the contractor of site arrangements, i.e. hours of work, services (air, water, etc.), welfare facilities that he can use (toilets, canteen, first aid), fire prevention arrangements, emergency procedures, etc.
- informing contractor of procedure for obtaining clearance to excavate and, where known, the routes of underground services.

Reciprocally, the contractor's responsibilities include:

- having a safety policy and ensuring it is implemented on the site
- preparing a programme for carrying out the contract
- preparing method statements for each phase of the work
- ensuring all relevant statutory legislation is complied with
- organizing the work to enable it to be carried out in a safe manner
- informing his own and sub-contractor's employees of special hazards on the site and training them in the precautions to be taken
- employing only skilled workmen and providing competent supervision
- informing the occupier of any equipment, materials or processes he may use that could give rise to a risk to the health of the occupier's employees and seek the occupier's permission before using them
- informing the enforcing authority of the intended commencement of the work
- taking all necessary steps to ensure that the occupier's employees are not put at risk from the manner in which the work is done
- providing all the equipment, materials and services needed to carry out the work
- instructing his and sub-contractor's employees in the site emergency procedures

- instructing his own and sub-contractor's employees of the site facilities they can use
- checking for underground services before excavating
- providing any necessary fire-fighting equipment for the part of the site under his control.

Within the above responsibilities, typical subject matters to be considered include:

- area of the premises over which contractor given control
- procedure for the transfer of control for that area
- means and routes of access
- use of site services; electricity, air, water, gas, etc.
- use of site welfare facilities; canteen, first aid, toilets, etc.
- site arrangements:
  - hours of work
  - access, traffic routes, speed limits
  - parking
  - emergency procedures
  - no-smoking areas
- site safety rules
- safe systems of work
- restrictions on use of occupier's equipment
- contractor's safety policy
- contractor's insurance cover and any restrictions or limitations
- liaison, communications and points of contact
- special training
- statutory requirements applicable to the site and the contract
- use of hazardous and dangerous materials, substances and equipment
- hand-over procedure at the end of the contract.

By working closely with contractors, the work can be expedited and the site made a safe place to work.

Sources of advice and guidance:

- HSE publications
  - HS(G) 46 'A guide for small contractors. Site safety and concrete construction' 1989 (ISBN 0 11 885475 5)
  - HS(G) 130 'Health and safety for small construction sites' 1995 (ISBN 0 7176 0806 9)
- 'Construction Hazard and Safety Handbook' by R W King and R W Hudson, Butterworth-Heinemann
- IOSH booklet no: 1, 'Safety with contractors in the motor industry'.

14.5 Access equipment

The relevant legislation is *The Construction (Health, Safety and Welfare) Regulations* which refer to *'supporting structures'*, i.e. any structure or device that supports a platform, and includes scaffolding, working platforms, access equipment, etc. However, since scaffolding is used so extensively it will be dealt with in some detail.

Scaffolds must:

- be erected by trained and competent person who is certificated by CITB or approved training organization

- be inspected by a competent person:
  - before being taken into use
  - every seven days
  - after inclement weather
  - after changes to the scaffolding
  - report of each inspection kept available in site office
- be of good construction, suitable and sound material and adequate strength
- not contain defective materials
- be securely supported, properly braced and, if relying on building for stability, rigidly attached to the building
- have standards or uprights vertical or slightly inclined towards the building with feet on adequate base plate
- have ledgers horizontal
- have well secured putlogs
- have hand rail 1 m high and toe boards 150 mm high on gangways, stairs and working platforms where risk of falling more than 2 m.
- gangways and working platforms to be at least 600 mm wide

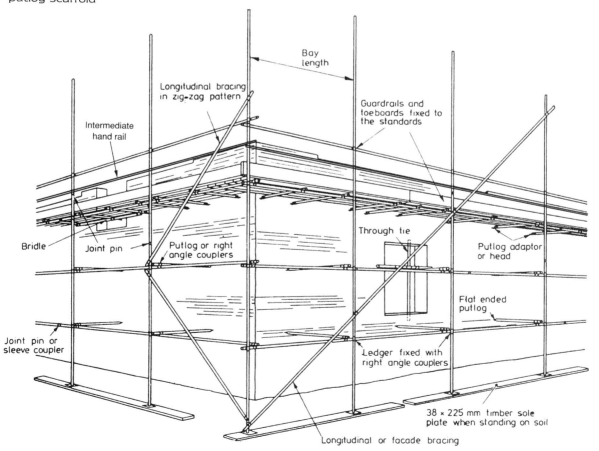

Figure 14.1 Typical putlog scaffold

... free standing towers must be stable ...

- scaffold boards to be laid so that:
  - no gaps large enough to cause injury
  - joins level so no risk of tripping
  - supported every 1 m for boards 32 mm thick
  - supported every 2.6 m for boards 50 mm thick
- access ladders to project at least 1 m above platform level and be properly secured
- if partially erected or dismantled, display a prominent warning notice and have access blocked
- be properly maintained and kept in good condition at all times
- timber parts of good quality wood, stripped of bark and left unpainted (so defects are not hidden).

Free-standing access towers must:

- be no higher than three times length of shortest side
- be constructed to be stable (use of cross braces)
- be used only on firm level and even surface
- have access ladders secured inside tower
- if mobile (fitted with wheels):
  - not be moved with anyone on the tower
  - be moved by applying force at or near the base
  - have wheels locked when in use.

'Snorkel' and 'bean stalk' type passenger lifts must:

- be used only on firm level ground
- use outriggers if provided
- follow maker's recommendations for load and radius
- not be overloaded.

Sources of information and guidance:

- HSE publications:
  - Guidance notes:
    - GS15 General access scaffolds
    - GS28 Safe erection of structures; pt 3 Working places and access
    - GS31 Safe use of ladders, step ladders and trestles
    - GS42 Tower scaffolds
    - PM28 Working platforms on fork lift trucks
    - PM30 Suspended access equipment.

14.6 Safety in demolition

Within the construction industry, which has a notoriously bad safety record, one of the more dangerous operations is that of demolition. However, the work can be carried out in relative safety if certain practices and rules are followed.

The legislation covering demolition is very brief and is contained in r.10 of *The Construction (Health, Safety and Welfare) Regulations 1996* which says:

Demolition or dismantling
10. (1) Suitable and sufficient steps shall be taken to ensure that the demolition or dismantling of any structure, or any part of any structure, being demolition or dismantling which gives rise to a risk of danger to any person, is planned and carried out in such a manner as to prevent, so far as is practicable, such danger.
(2) Demolition or dismantling to which paragraph (1) applies shall be planned and carried out only under the supervision of a competent person.

It should be noted that the requirement is to *prevent, so far as is practicable, such danger*, i.e. the question of reasonable practicability and cost of taking suitable precautions does not enter the consideration.

The demolition work is subject to CDM and as such has to be properly planned and executed. It can conveniently be considered in three phases:

1. Planning the work
2. Health hazards
3. Working methods.

14.6.1 Planning the work

In the initial stages, consideration to be given to:

- the type of structure or building
- its condition
- the existence of services – whether live or dead
- presence of residual hazardous substances from previous use, such as asbestos, flammable and toxic materials
- effect of or on adjacent buildings
- access to site for plant
- storage of demolished materials
- removal of debris and spoil
- programme of work:
  - sequence of working

- methods of work:
  - prepare method statements
  - systems of work including permits-to-work
- selection of contractor:
  - competent
  - experienced in the type of demolition
  - provide competent supervision.

14.6.2 Health hazards

Hazards to health can arise from the presence of:

- gases, fumes, vapours and dusts which may be asphyxiant, toxic, corrosive or carcinogenic and cause ill-health through:
  - inhalation
  - ingestion
  - absorption through the skin or open wounds
- flammable liquids and materials.

Typical substances met in demolition include:

- lead
- asbestos
- cement dust
- silica
- residues from previous processes on site
  - manufacture, use, handling, storage, etc. of known or unknown materials.

Appropriate precautions must be taken and personal protective equipment provided.

14.6.3 Demolition methods

Methods of demolition include:

- Hand demolition:
  - follows reverse order from construction
  - dangers from falling debris so no working allowed under parts being demolished
  - poor access over rubble and to parts to be demolished
  - lack of working platforms at parts being demolished
  - if chimney used as debris chute, clear base regularly
- Mechanical demolition
  - demolition ball:
    * vertically
    * swinging in line with jib   } heavy duty crane must be used
    * slewing jib
  - pusher arm
  - hydraulic hammer
  - power grapple and shears
  - grabs
- Where mechanical demolition used:
  - roof trusses hand-demolished to bearer plate
  - hand demolition to 1 m wide from adjacent buildings
  - stability of adjacent building to be checked and maintained

- only machine and banksman within 6 m of part of building being demolished
- machine should have clear working space of 6 m from building
- the building should not be entered once demolition has started

- Deliberate collapse
  - structure pre-weakened then use either explosives or wire rope pulling.

Note: this technique requires knowledge of the strength of building structures and should only be carried out by competent operatives under strict supervision.

- Explosives
  - to be used only by specialists with knowledge of:
    * firing charges and sizes
    * blast protection necessary
    * area of exclusion zone
    * firing programme
    * how to handle misfires
    * precautions to prevent predetonation from radio and similar interference
    * warning signs and policing of exclusion zone
    * security arrangements for storage of explosives

- Wire-rope pulling
  - not to be used on structures over 21 m high
  - steel rope with minimum diameter of 38 mm to be used
  - winch or pulling vehicle positioned at least two times building-height from building
  - winch or pulling vehicle to have steel cab or cage to protect operator
  - exclusion zone to extend from building and on each side of the rope a distance at least three-quarter of the distance of winch or pulling vehicle from building
  - if pulling fails:
    * building or structure must not be approached or entered
    * other mechanical means to be used to complete demolition.

Fire should **never** be used as a method of demolition.
Special techniques that should be described in the work method statement are required for:

- pitched roofs
- filler joist floors
- floors supported by steel beams
- jack arches
- reinforced concrete floors
- framed structures
- in-situ cast reinforced concrete:
  - beams
  - columns
  - walls
  - suspended floors and roofs
- pre-cast concrete structures, floors and wall panels
- pre-stressed concrete

Demolition is a highly dangerous process and should only be undertaken by specialists with the appropriate qualifications and experience.

Guidance can be obtained from:

- BS 6187, Code of Practice – Demolition
- HSE Guidance Notes: GS29 Health and safety in demolition work

14.7 Safety with excavations

Many building and construction projects require excavations, from a simple trench for pipe-laying to deep foundations for a major building. Any work at levels below the surface of the ground can give rise to risks from collapse or slippage of the ground, falls of persons or materials, and flooding. By following simple but well established techniques, those risks can be reduced to a minimum.

This section is not concerned with tunnelling which is a specialized operation that should be carried out only by organizations competent and experienced in that type of work.

It is accepted that excavations less than 1.2 m (4 ft) deep do not present a serious risk to those working in them. In any excavation of greater depth suitable precautions must be taken.

Typical hazards from excavations:

- collapse of sides
- materials falling onto those working in the excavation
- people or vehicles falling in
- undermining adjacent structures and buildings
- damage to underground services
- asphyxiating fumes and gases
- flooding.

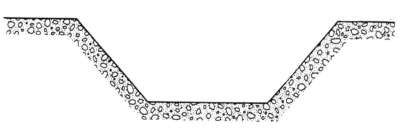

(a) Battering the sides

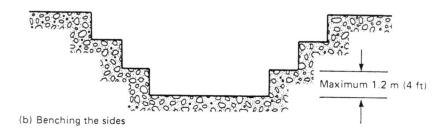

(b) Benching the sides

Figure 14.2 Safe trenching methods without the use of timber

14.7.1 Prevention

Considering each of the hazards listed above prevention methods include:

- Collapse
  - benching
  - battering
  - shoring.

  If shoring:
  - timber to be used must be:
    * inspected
    * in good condition
  - be erected by experienced workmen under the control of a competent person experienced in the techniques
  - be inspected before excavation is used
  - can also be:
    * trench boxes with hydraulic struts
    * metal-sheet piling
- Material falling in
  - store any spoil, plant or materials clear of the edge of excavation
  - fit toe boards along edge of opening to prevent loose material falling in
- People falling in
  - if greater than 2 m (6 ft 6 ins) deep:
    * provide substantial barrier
  - if public can approach:
    * fence off completely
- Vehicles falling in
  - provide suitable barrier or stop-log which should be painted to stand out
  - if need to tip into excavation provide stop-blocks at appropriate distance from edge
- Adjacent structures
  - check foundation or footing of structure
  - if risk of weakening them shore the walls before digging
- Services:
  - include gas, water, electricity, telephone
  - establish runs:
    * from local plans
    * by using pipe/cable locator (Note: these devices do not detect plastic pipes)
    * hand-dig trial holes to confirm location of services
    * hand-dig in vicinity of services
    * when exposed provide support for service carrier
    * report to service provider any damage to service
- Gases and fumes
  - from engines, boilers, LPG cylinders, solvents, etc.
  - check atmosphere in excavation before each shift
  - duct fumes away from excavation, or
  - provide adequate fresh air ventilation
  - fumes can lie in excavation for some time after source removed
- Flooding
  - provide suitable ladders for escape
  - provide pump of adequate capacity to keep excavation dry.

Inspections play a critical role in accident prevention. Inspections should occur for:

- all excavations:
  - before being used
  - every day when in use
- excavations deeper than 1.98 m (6 ft):
  - at the beginning of every shift.

These inspections do not have to be recorded but it is prudent to do so. However, the following inspections of all excavations must be recorded on Form 91:

- after explosive charges have been fired
- after damage to shoring
- after fall of earth or collapse of material
- every seven days.

The precautions outlined above can be implemented with very little disturbance to the work but could prevent the enormous amount of disruption that inevitably follows an accident.

15   Manual handling

Over one third of all reported lost-time absences are attributed to injuries caused by manual handling. This problem is common across Europe and has resulted in a Directive aimed at reducing this toll. In the UK the contents of the Directive are contained in *The Manual Handling Operations Regulations 1992* (MHOR).

A number of practical techniques have been developed to facilitate the manual handling of loads and hence reduce the toll of injuries and some of these are considered in this chapter.

15.1 The Manual Handling Operations Regulations 1992

These Regulations are aimed at reducing the appalling toll of injuries from manual handling. The requirements of the Regulations are summarized below.

r.2   Defines:

- manual handling operations as *any transporting or supporting of a load (including lifting, putting down, pushing, pulling, carrying, or moving thereof) by hand or by bodily force*
- load as including any person and any animal.

Puts duties on employers regarding their employees; the self-employed are responsible for themselves.

r.3   The Regulations do not apply to ship-board activities of sea going ships.

r.4   Employers should avoid manual-handling operations that could put employees at risk of injury [i.e. use mechanical handling wherever possible].

Where manual handling is unavoidable, employers should:

- carry out a risk assessment of the handling operation covering:
  - the tasks
  - the loads
  - the working environment
  - capacity of the individual to do the work
  - other factors such as the effect of protective clothing, etc.

> - reduce the risks from manual handling to a minimum
> - provide information on:
>   - the weight of loads to be handled
>   - off-centre and out-of-balance loads
> - review the risk assessment when conditions change.
>
> r.5 Employees are required to follow instructions aimed at reducing the risk from manual handling.

Under HSW, employers are required to provide suitable training to employees and this extends to include training in the various manual handling techniques – see Section 15.2.

15.2 Safe manual handling

Of the enormous toll of injuries attributed to manual handling by far the greater number (two-thirds) are strains and sprains. By knowing the mechanics of the body as they apply to lifting and handling and by following some simple rules the toll of these accidents can be much reduced.

15.2.1 Definitions

- **Manual handling** – any transporting or supporting of a load by hand or bodily force whether lifting, putting down, pushing, pulling, carrying or moving.
- **Sprain** – severe wrench or twist of a ligament or muscle of a joint causing pain and swelling of the part resulting in pain and difficulty in moving.
- **Strain** – injury done to a limb or part of the body through being forcibly stretched beyond its proper length.

15.2.2 The lifting mechanics of the body

Understanding the mechanics of the parts of the body that are involved when a load is lifted or carried is basic to developing techniques and practices to ensure the muscles are not overloaded.

Figure 15.1 shows a simple line diagram of the skeleton carrying a load (W) at a distance (y) from the spine.

Figure 15.2 shows a diagrammatic detail of the vertebrae of the spine with the discs, the spinal cord and the back muscles attached to a horn-like projection from each vertebrate. The backbone pivots about the discs between the vertebrae with reactionary force applied by the muscles. The whole of the load that is lifted is taken by the spine.

The bending moment about the spine from the load $= W \times y$

This is resisted by the moment of the pull of the muscle times its distance from the vertebrate $= P \times r$

Therefore for balance $W \times y = P \times r$

Thus the load on the spinal muscle $P = W \times y \div r$

If r is small compared to y then the load on the spinal muscle is many times greater than the load being carried. For example if a load of 10 kg is carried at elbow's length, i.e 400 mm from the spine, and the distance of the spinal

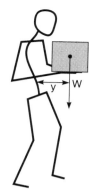

Figure 15.1 Body mechanics of lifting

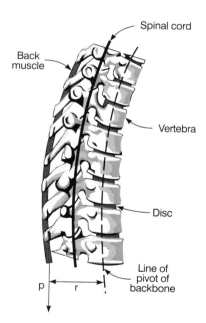

Figure 15.2 Mechanics of the backbone

muscle from the centre of the disc is 20 mm then the load on the spinal muscle is:

10 kg × 400 mm ÷ 20 mm = 200 kg (450 lbs or 4 cwts)

Thus to prevent back strain the load carried should be:

- kept as low as possible
- carried as near to the body as possible.

Other muscles are attached to bones near to their pivot point, in the arms and wrists for example, and similarly carry many times the load of the weight being lifted.

Figure 15.3 shows recommendations for maximum loads at various distances from the body and at differing heights.

212 Manual handling

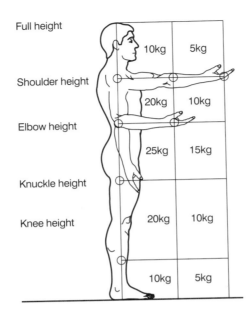

Figure 15.3 Suggested maximum loads at various distances from body

15.2.3 Preventing strain

Where a load has to be moved there are a number of simple steps that can be taken to prevent back and other muscle strain:

- wherever possible the load should be moved by mechanical means
- if not, an assessment made of the risks involved in manually handling it
- the load should be split down into manageable sizes
- each part should be within the lifting capability of the employee
- if still too heavy, assistance should be sought to help with the lift
- employees should be trained to:
  - follow the correct lifting techniques including kinetic handling by:
    * placing feet firmly
    * gripping load firmly
    * using arms and legs to lift – raising load in stages, resting it on knees or thighs while the position of the back is changed
    * not trying to lift with the back, i.e. not changing position of the back when lifting or holding the load
    * taking the strain slowly when picking up the load
    * not snatching the load
    * holding the load as close to the body as possible
    * not twisting the body when lifting or carrying a load
  - estimate the weight of loads
  - ask for assistance if the load is too heavy
  - wear appropriate protective clothing:
    * gloves to protect hands from splinters
    * safety shoes to protect feet should load be dropped
  - report any difficulties to the supervisor.

When moving loads on a trolley:

- Pulling
  - the whole strain is taken by the back muscles – see Figure 15.4
  - if load moves suddenly it can run on and injure employee's feet

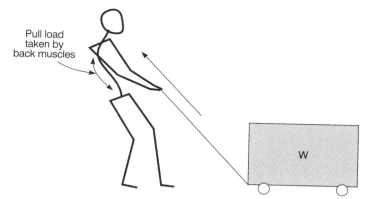

Figure 15.4 Dynamics of pulling a load

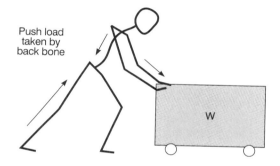

Figure 15.5 Dynamics of pushing a load

- Pushing
  - at optimum height of push (about mid-chest height) the strain passes down the backbone – see Figure 15.5
  - pushing at too high a level puts excessive strain on stomach muscles and can cause strain of shoulder muscles and hernia
  - pushing at too low a level is ineffective.

Typical problems encountered with loads include:

- weight
- shape
- size
- compactness or looseness of packing
- off-centre loads.

Where continued repetitive movement or handling causes pain – called repetitive strain injury (RSI) or work related upper limb disorder (WRULD) – the rhythm or routine of the work should be changed or arrangements made for a rotation between different jobs. If condition persists seek medical advice.

For further guidance, consult HSE publication no. L23 'Manual handling, guidance on the Regulations', which gives advice on the interpretation of the Regulations and examples of a number of techniques that can be used.

16   Mechanical handling

Mechanical handling refers to any powered means for moving or lifting loads. The more common means are:

- lifts, including hoists
- cranes, including lifting tackle
- conveyors
- powered trucks, including driverless trucks.

All items of handling equipment, whether powered or manually operated, must comply with:

- for equipment purchased for use before 31 December 1992:
  - *The Provision and Use of Work Equipment Regulations 1992* (PUWER)
- for new equipment purchased on or after 1 January 1993:
  - *The Supply of Machinery (Safety) Regulations 1992* (SMSR) and subsequent amendments including *The Lift Regulations 1997*.

All lifting equipment should:

- be tested before being put to work for the first time
- have a certificate issued
- be examined regularly
- be properly maintained.

The different types of handling equipment are considered in more detail in this chapter.

16.1 Lifts

The current position regarding the construction and use of lifts has changed recently with the coming into effect of new regulations *The Lift Regulations 1997* incorporating the content of an EC Directive (no: 95/16/EC) *on the approximation of the laws of Member States relating to lifts*. This is an Article 100A Directive concerned with supply and aimed at removing barriers to trade. It is concerned only with the construction of lifts and does not concern itself with the use, maintenance or periodic inspections. Those requirements are contained in sections of the *Factories Act 1961* (FA) that remain extant. However, that too is likely to change since the HSC is considering new regulations concerning the safe operation of lifts.

This section deals with the construction and supply of lifts and covers both the existing FA requirements and those of the new regulations. Under the new regulations, there is a run-in period to 30 June 1999 during which lifts can be manufactured to the requirements of either the FA or the new regulations. After that date all new lifts must conform to the new regulations. Both these requirements are summarized below.

A **lift** is defined:

- in the *Factories Act*, as having *a platform or cage the direction of movement of which is restricted by a guide or guides.*
- in the new Regulations, as being *an appliance serving specific levels, having a car moving:*
  (a) *along guides that are rigid; or*
  (b) *along a fixed course even where it does not move along guides which are rigid (for example a scissor lift) and*
  (c) *inclined at an angle of more than 15 degrees to the horizontal and intended for the transport of:*
  - *persons*
  - *persons and goods*
  - *goods alone if the car is accessible, that is to say, a person may enter it without difficulty, and fitted with controls situated inside the car or within reach of a person inside.*

As regards the safe *use* of lifts and hoists, the requirements contained in FA still apply.

> s.22 Requires that all lifts and hoists must:
>
> - be of good mechanical construction, sound material and adequate strength
> - be properly maintained
> - be examined by a competent person every six months and results recorded (these examinations are normally carried out by the company that insure the equipment)
> - if examination shows repairs needed, a copy of report must be sent to HSE
> - have their liftway/hoistway fully enclosed and fitted with gates at landings
> - enclosure to be such that neither person or goods can be trapped between the cage and fixed parts or counterbalance weights in the shaftway
> - gates must be interlocked so they cannot be opened until the cage is at the landing and when the gates are open the cage cannot be moved
> - cage to be marked with safe working load.
>
> s.23 Requires lifts for carrying persons to include:
>
> - automatic device to prevent the cage from over-running (the cage floor should stop level with the landing floor)
> - cage doors must be interlocked in the same manner as the landing gates
> - cage fitted with dual ropes, each capable of sustaining full load
> - cage to have arrester device capable of bringing a fully loaded cage to rest if the rope breaks.

Continuous operating lifts of the paternoster type are exempt from having gates but must have edge trips to stop movement in the event of anyone becoming trapped between part of the cage and the cill or lintel of a landing. Their speed of movement is restricted.

Escalators and travelators, while strictly not lifts, transfer people between fixed points. Precautions to include:

- skirting to prevent trapping between tread and balustrading
- combs to tread at entry and exit
- adequate clear area of floor at entry and exit
- clearly indicated emergency stops devices
- regular and proper maintenance including cleaning the area of the return track
- periodic thorough examination.

There are a considerable number of exemptions for special purpose lifts which are detailed in *The Hoists Exemption Order 1962*.

With scissor lifts:

- access must be prevented to the scissor mechanism by:
  - setting the base in a pit
  - having roller screens, shutters or similar between the table and base wherever mechanism is exposed
  - providing fences round the non-operating sides.

When servicing a scissor lift, the table should be retained in the elevated position by chocking the moving wheels in the base. Do *not*, repeat *not*, attempt to chock the table since this can tilt upwards and allow the scissor mechanism to close.

New lifts put on the market are required to conform with the *Lifts Regulations 1997*, the contents of which are summarized below.

r.3　The Regulations apply to:

- lifts permanently serving buildings and constructions
- safety components for lifts.

r.4　They do not apply to:

- cableways
- military or police lifts
- mine winding gear
- theatre elevators
- lifts in means of transport
- lifts that are part of machinery
- rack and pinion trains
- construction site hoists

r.5　• lifts purchased before 1 July 1997

r.6　• lifts purchased before 30 June 1999 that comply with the FA

r.7　• lifts to which other EU Directives apply.

r.8,9　All lifts and safety components must:

- satisfy the *essential safety requirements* (ESRs) through conformity with EN standards
- follow conformity assessment procedure
- carry the CE mark
- be accompanied by a declaration of conformity
- be safe.

r.10 Suppliers who are not manufacturers of lifts or safety components must ensure lifts they supply are safe.

r.11
- Person in charge of a building and the lift installer to keep each other informed to ensure safe and proper use of the lift.
- No unnecessary pipework or wiring to be installed in lift shaft.
- Designer to provide manufacturer and installer with all information necessary to ensure safe operation.
- The declaration of conformity must be kept for ten years.

r.12 These conditions need not be complied with if the lift is:

- destined for market outside EU
- at an exhibition when:
  - a notice stating it does not comply must be displayed
  - visitors are adequately protected.

r.13 Conformity assessment procedure for lift to include:

- final inspection
- manufacture to appropriate quality assurance scheme
- unit verification.

Conformity assessment procedure for a safety component to include:

- EU type examination
- manufacture to full quality assurance scheme.

r.14 Supplier who puts lifts and safety components on the market must ensure they conform.

r.15 Conformity assessment procedures must be carried by a *notified body*.

r.16 *Notified bodies* are appointed by the Secretary of State.

r.17 Notified bodies can charge fees for carrying out conformity assessment procedures.

r.18 A lift or safety component bearing the CE mark presumes conformity.

218 Mechanical handling

16.1.1
Essential health and safety requirements (ESR)

Points to be considered include:

- General:

  1. ESRs of Framework Directive apply
  2. safety of car
  3. means of suspension and support
  4. control of loading
  5. machinery
  6. controls.

- Hazards to persons outside the car

  1. access to lift shaft only if lift isolated
  2. suitable protection at ends of lift travel
  3. gate interlocks must be effective.

- Hazards to persons in the car:

  1. cars doors must remain shut except at landings
  2. car must be fitted with fall arrester
  3. buffers provided at bottom of shaftway
  4. car must not move if arrester device not operational.

- Other hazards:

  1. motorized car and landing doors must not crush when closing
  2. landing doors must be fire-resistant
  3. counter weights to be clear of car or other obstruction
  4. cars to have:
     - means of escape for trapped persons
     - two-way communication system
     - arrangements that allow current operation to be completed if temperature gets too high
     - adequate ventilation
     - emergency lighting
     - communication and lighting to work in event of power failure
  5. lifts capable of use in emergency must have priority controls for emergency service use.

- Marking:

  1. each car must be marked with maximum load
  2. car to contain instructions for emergency escape.

- Instructions for use:

  1. of safety components to be in language of installer
  2. for operation in language of user.

Many of the requirements of these Regulations reflect the requirements of the FA and good UK practice but there is greater emphasis on manufacturing documentation and quality assurance systems.

16.2 Cranes

Cranes are lifting equipment that have no restraint on the direction of movement. Typical cranes are:

- chain blocks — single direction of movement
- mono rails — two directions of movement
- overhead travelling cranes
  mobile cranes
  tower cranes
  jib cranes
  } — three directions of movement.

The construction of all new cranes bought since 1 January 1993 should comply with the *Supply of Machinery (Safety) Regulations 1992* and carry the EU mark 'CE'. Those Regulations refer to:

- *lifting machinery* rather than cranes
- *lifting accessories* – meaning components or equipment not attached to the machine and placed between the machinery and the load or on the load in order to attach it to the crane hook
- *separate lifting accessories* – meaning accessories that help to make up or use a slinging device, such as eyehooks, shackles, rings, eyebolts, etc.

In use, cranes and lifting tackle should still comply with the requirements of ss. 26 and 27 of FA.

> s.26 Refers to lifting tackle as the equipment used to join the load to the crane and includes:
>
> - chain slings
> - rope slings
> - rings
> - hooks other than a hook that is part of a crane
> - shackles
> - swivels.
>
> Requires that:
>
> - it is examined by a competent person every six months
> - it is marked with its safe working load or that information is kept immediately available, i.e. in the tackle store
> - it is not used above its safe working load
> - except for fibre ropes and slings, it is tested before being used for the first time
> - fibre ropes and slings should have a test certificate of a typical sample of the rope
> - it is properly stored on rack or suitable shelving
> - wrought iron chains should be heat treated every fourteen months. Note: modern chains of high tensile steel must *not* be heat-treated (which may reduce the strength of the chain).

s. 27 Refers to cranes as lifting machines, and includes:

- crab
- winch
- teagle
- pulley and chain blocks
- gin wheel
- transporter and runway.

Requires that all cranes of whatever type must:

- be of good construction, sound material and adequate strength
- have a test and examination certificate
- be properly maintained
- be examined by a competent person every fourteen months and a report made
- if examination shows repairs needed a copy of report to be sent to HSE
- be clearly marked with safe working load which must not be exceeded
- where cranes run on a track:
  - not approach nearer than 6 m (20 ft) to anyone working on the track
  - have rails of proper size, adequate strength and have even running surface
- where anyone working above ground level could be struck by the crane or its load, they must be warned of its approach.

To complement these requirements and ensure safe use of cranes:

- driver, slinger and banksman must be trained and certificated
- driver should be assisted by a banksman who directs the driver in the movement of loads
- there should be only one banksman to direct crane driver
- a slinger, or if no slinger the banksman, slings or attaches the load to the crane hook
- lifting equipment and tackle must be properly stored
- there should be regular visual inspections of ropes and rope slings – this should be done every time they are used
- all standard hooks should have a safety catch – some 'C' hooks, specially designed to prevent displacement of the load without a safety catch, are still in use
- overload alarms should be fitted wherever feasible
- cut-out switch should be fitted between the sheaves of the lifting hook and crane crab to prevent jamming and over-straining of the lifting ropes (it is expensive to replace a lifting rope).

Additional precautions needed when using mobile cranes include:

- use of outriggers
- ensure outrigger feet are:
  - on solid base boards
  - on well compacted level ground

Mechanical handling **221**

- employ only trained, certificated and authorized drivers
- allow room for swing of counterbalance weight particularly when near buildings
- if working near overhead power lines, position warning flags or bunting and post warning notices
- if crane hired-in, check:
  - examination and test certificate is up to date – if not, do not allow on site
  - driver supplied by hirer is fully trained and certificated by CITB – check licence
  - the crane is suitable for the work to be done
  - the hirer has adequate insurance cover (min £1m), either general public liability (PL) or special cover for the job.

Construction site lifting equipment is subject to the above requirements and, because of the hostile environment in which they have to work, may require more frequent examinations and checks.

16.3 Conveyors

Conveyors are used to carry a flow of goods or materials, and can present a number of hazards. There are two broad types of conveyor: roller conveyors and belt conveyors. A common hazard is that of trapping, either between adjacent rollers or under the belt.

16.3.1 Roller conveyors

- free running type:
  - normally gravity feed
  - no trapping hazards except at entry from belt conveyor (Note: in Figure 16.1 the first roller C is arranged to lift out of its housing)
- power-driven rollers:
  - if all rollers driven there are no intake traps
  - if alternate rollers driven, trap can be formed between driven and idler rollers – provide strip guard across conveyor at these points

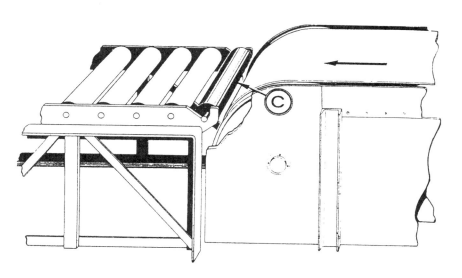

Figure 16.1 Preventing free running roller trap

222 Mechanical handling

16.3.2
Flat-belt and trough conveyors

- guarding to be provided at intakes to:
  - belt drive from the motor
  - head pulley for at least 1 m from intake
  - tail pulley for at least 1 m from intake
  - snub or tensioning pulleys
  - rollers where belt changes direction
  - rollers under feed hoppers where side of hopper prevents belt lifting off rollers
  - return belts below conveyor to provide protection when sweeping floor

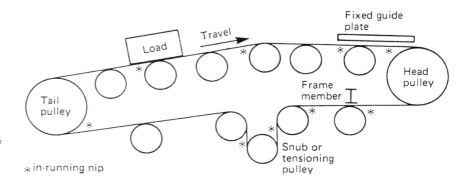

Figure 16.2
Diagrammatic layout of belt conveyor showing in-running nips

* in-running nip

- guarding of long straight runs can be by trip wire which must:
  - have a lock-out switch at each end
  - be within easy reach of anyone trapped by belt
  - be arranged so that pulling in either direction trips the conveyor
  - when tripped, have controls arranged to require reset before conveyor can be restarted
  - keep switches clean and clear of debris
  - have its function checked regularly
- guarding by distance fencing must:
  - be high enough and positioned so belt cannot be reached
  - have interlocked gates to permit access for adjustment of, repairs to and cleaning of the conveyor
  - provide reasonable gap under fence to allow for sweeping the floor

With storage systems based on areas of flat or gravity roller conveyors, adequate longitudinal and cross walkways must be provided to enable operators to get to loads and push them safely (see Figure 16.3).

16.4 Powered trucks

This section does not cover road vehicles but only those trucks used solely within a works premises.

A common requirement for the safe use of all powered trucks is that the drivers are fully trained and competent.

16.4.1
Training

The training of drivers should:

- be by qualified instructors – check their credentials. If a training centre they should be members of the Industrial Truck Training Assocation
- be held in-house or at accredited training centre

Mechanical handling 223

Figure 16.3 Walkways between and across roller conveyors

- cover:
  - driving theory, especially for fork trucks with rear-wheel steering and other specialized trucks with multi-wheel steering
  - use of basic controls
  - practical driving techniques
  - practical experience in the workplace
  - test of theoretical knowledge and practical competence.

Drivers who successfully complete training should be given a certificate of competence (licence).

16.4.2 Fork-lift trucks

Candidates for fork-lift truck training should:

- be over eighteen years of age
- have a medical check before training starts
- have stereoscopic vision. If monocular (one-eyed) they should be referred for a medical check on their vision
- not be colour-blind
- if epilectic or diabetic, be referred to doctor for medical check
- not be known drug users.

Safe operating techniques for fork lift trucks include:

- trucks should be fitted with back and overhead guards
- driver should be aware of the outward swing of the back of the truck when turning due to back-wheel steering

- powered steering prevents spinning of steering wheel if road wheel hits an obstruction – if manual steering the steering wheel should be recessed for the steering knob
- carrying of passengers prohibited unless purpose-built seat fitted
- trucks must not be used to raise persons unless fitted with a special-purpose working platform, for details see HSE guidance note PM28
- floor must be in good condition and level
- when travelling up or down a slope the load must be towards the top of the slope
- must not be driven with the forks elevated, either loaded or unloaded
- ignition key or starter card must not be left in an unattended truck
- priority right of way must be given to pedestrians
- loads must be within makers recommended limits
- lifting chains should be examined every six months
- pedestrian accessways should be separate from truck lanes
- when charging battery electric trucks:
  - hydrogen given off by battery
  - battery cover should be removed or perforated cover used to allow hydrogen to escape
  - high-level ventilation should be provided to charging area to disperse hydrogen
  - provide water supply and eye-wash bottles in case of acid splash
  - when replacing batteries ensure lifting hooks do not make contact with a battery terminal.

Additional special training should be given to drivers of rough terrain lift trucks.

Most common causes of accidents with lift trucks are:

- turning at speed
- turning across a slope
- turning with the load elevated
- skylarking by unsupervised youngsters.

Drivers guilty of any of these practices should have their authorization to drive (licence or certificate) suspended until further training completed.

For guidance on the safe use of lift trucks:

- British Industrial Truck Association's booklet 'Operator' Safety Code'
- HSE guidance booklet no. HS(G) 6 'Safety in working with lift trucks'
- HSE guidance booklet no. HS(G) 113 'Lift trucks in hazardous areas'
- HSE guidance booklet no. HS(G) 136 'Workplace transport safety. Guidance for employers'
- HSC Approved Code of Practice COP 26 'Rider-operated lift trucks – operator training'
- HSE guidance note no. PM 28 'Working platforms for fork lift trucks'.

16.4.3 Pedestrian operated trucks

Usually the operator walks in front of the truck where the main hazard is over-run of the vehicle with consequent injury to driver's feet.

Precautions include:

- over-run trip on control handle
- fitting of 'cow catcher' guard to front wheels
- wearing of safety footwear.

16.4.4 Driverless vehicles (robot tugs)

Used where frequent transfers made between discrete fixed points. Normally follow a fixed route with guide wire let into the floor and travel at a controlled speed. Main precaution is the provision of a trip device fixed to the front of the towing truck such that the truck with full load will stop before the main body of the truck reaches any obstruction it meets.

16.4.5 Road vehicles

Where road vehicles are used solely for transport around a site they should be:

- maintained to *Road Vehicles (Construction and Use) Regulations 1986* standards
- driven in accordance with the Highway Code
- have separate entrances into buildings from pedestrians
- use a banksman or audible warning when reversing.

Site roads should:

- conform to standards for road traffic
- be well defined and separate from pedestrian footpaths
- be maintained in good condition
- be provided with guard rails where pedestrian exits from buildings let straight onto a roadway
- have speed limits [between 15 and 20 mph for medium-sized site]
- have 'sleeping policemen' to restrict speeds.

Essential ingredients for the safe use of powered vehicles on a site are:

- the driver's awareness of manoeuvres that could put others at risk
- the driver's taking action to check that the way is clear for the vehicle to move, whether forwards or backwards
- giving priority to pedestrians.

Various types of powered vehicles can be used safely around sites, but the degree of safety is determined by the driver's skill and consideration for others.

17 Safe use of electricity

Electrical power is the most convenient and flexible source of power in both industry and in the home. It is also probably the most dangerous because:

- you cannot see it
- you cannot smell it
- if you feel it, it may be too late.

But it can be used safely if its dangers are understood, certain simple rules are followed and it is treated with respect.

Legislation lays down basic requirements to be met in the use of electricity but there are, in addition, a number of well tried techniques that should also be followed. Both these aspects are dealt with in this chapter.

17.1 The Electricity at Work Regulations 1989

The regulations relating to the safe use of electricity at work refer to all work situations, this includes mines and quarries where there are particular and specific requirements. This section deals only with the requirements for the safe use of electricity in normal industrial and commercial workplaces.

17.1.1 Legislation

- The *Electricity at Work Regulations 1989* supported by HSE's publication no. HS(R)25 'Memorandum of Guidance on the Electricity at Work Regulation 1989'.
- Regulations 14–19 of the *Provision and Use of Work Equipment Regulations 1992* (PUWER).
- Although not legislation, the Institution of Electrical Engineer's Regulations for Electrical Installations (The IEE Wiring Regulations) now incorporated in BS 7611 lays down standards for electrical installations up to 1000 volts ac that are recognized as being in compliance with statutory requirements.

Although made under the HSW, which requires that standards have to be met *so far as is reasonably practicable*, because of the dangers inherent in the use of electricity certain of the regulations are 'absolute' requirements, i.e. they must be complied with regardless of cost. However, it is recognized that if employers have taken *all reasonable steps and exercised all due diligence to avoid committing an offence* they will have fulfilled this absolute duty, but it will be up to them to prove it if an inspector queries it.

Safe use of electricity

The regulations to be complied with *so far as is reasonably practicable* are the following:

r.2 Defines:

- **conductor** as carrying electrical energy
- **danger** as risk of injury
- **electrical equipment** as anything that uses, generates, transforms, conducts, stores, measures, etc., electrical energy
- **injury** a death or personal injury caused by electricity
- **system** as including all electrical equipment deriving power from a common source.

r.3 Duties are placed on:

- employers and the self-employed to comply
- employees to:
  - co-operate with the employer in complying
  - comply themselves.

r.4 To prevent danger:

- all systems shall:
  - be properly constructed
  - be properly maintained
- any work on a system shall be carried out in a safe manner.

r.6 Electrical equipment working in a hostile environment must be designed and built to resist it.

r.7 All conductors in a system must:

- be insulated and
- be protected or
- have precautions taken to prevent danger.

rr.17–28 Relate to mines and quarries and are not covered by this part.

r.29 Lists those regulations that are absolute but allow a defence of *having taken all reasonable steps and exercised all due diligence*.

Those *absolute* regulations that must be complied with are:

r.4 part Protective equipment provided shall be:

- suitable for its intended use
- well maintained
- properly used.

r.5 No electrical equipment to be put into use if overloading its strength and capacity causes danger.

r.8 Where there may be danger from a conductor it shall be earthed or protected by other means.

r.9 Earth conductors must not contain switches or other means that could break the connection.

r.10 Every joint and connection must be sound.

r.11 Excess current protection must be provided to prevent danger.

r.12 Electrical equipment must be provided with:

- means to cut off the supply of electrical energy
- an isolating switch that can be secured (locked).

Generating equipment is excluded but must take precautions to prevent danger.

r.13 Where equipment has been made dead for working on, precautions must be taken to ensure it remains dead.

r.14 No one may work on a live uninsulated conductor unless:

- it is not reasonable to make it dead
- it is reasonable for the work to be done live
- suitable precautions have been taken, including the provision of protective equipment.

r.15 When working on or near electrical equipment there must be:

- adequate working space
- adequate access
- adequate lighting.

r.16 Where technical knowledge is necessary to prevent danger, the person working must:

- possess that knowledge or
- be experienced or
- be adequately supervised.

r.14 In addition PUWER requires that work equipment must be:

- provided with controls for:
  - starting and re-starting
  - changing operating conditions without danger.

r.15 • Stopping without danger and, if necessary, isolating power supplies.

r.16 • Emergency stopping which over-rides all other controls.

> r.17 Controls must be:
>
> - clearly visible
> - identifiable with marking if necessary
> - positioned so they can be operated safely
> - positioned so operator can see all parts of equipment; if not, controls must include audible warning and delay of start facility.
>
> r.18 Requires that control systems must:
>
> - be safe
> - not create danger in operation
> - not allow faults to create danger
> - accommodate power failures safely
> - not interfere with stop and emergency-stop control.
>
> r.19 Controls must include means of isolation from sources of energy which must be:
>
> - clearly identified
> - readily accessible (and reachable).

These Regulations lay down basic requirements for the installation and use of electricity on which can be built working practices that enable full use to be made of this source of power.

17.2 Safe use of electricity

Remember – electricity is lethal:

- you can't see it
- you can't smell it
- if you feel it, it may be too late.

When considering the safe use of electricity a number of terms arise that have particular meanings. Those terms include:

- **charged**
  - has acquired an electric charge:
    * through being connected to a live conductor
    * by electrical induction
    * of static electricity
- **competence**
  - having:
    * adequate technical knowledge
    * adequate experience of the work involved
    * a detailed knowledge of the particular process or equipment
  - so as to be able to:
    * recognize defects
    * assess their importance
    * recommend any necessary remedial action

- **dead**
  - carrying no charge, disconnected from all sources of electricity and connected to earth
- **earthing**
  - connecting direct to the mass of the earth so as to prevent any charge building up or being acquired. In earthed circuits the connection from the appliance to earth must be *solid* and not pass through any means of breaking the connection, i.e. switches
- **excess current protection**
  - means to prevent a circuit or appliance being subjected to a current beyond its capacity, usually either a fuse or circuit-breaker set to operate at a predetermined current level
- **double insulation**
  - the provision of two separate layers of insulation between the live parts and the part being handled. Double insulated appliances do not need to have an earth connection
- **duty holder**
  - a person who owes a duty. Under the *Electricity at Work Regulations 1989* this includes anyone who is in control of plant or equipment which, directly or indirectly, can cause danger or injury to other persons
- **insulation**
  - the protection provided on a conductor to prevent it making contact with another conductor, with earth or being touched by a person. The insulation may also have to provide protection against mechanical damage to the conductor
- **IP rating**
  - an international system for classifying levels of protection of equipment against the ingress of dusts and moisture. (IP = Index of Protection – see BS EN 60529 for details)
- **isolation**
  - the switching off of electrical supplies and locking the isolating switch in the OFF position
- **live**
  - being at a voltage through being connected to a source of electricity, i.e. connected to a conductor
- **electrical protection**
  - of plant and equipment:
    * by the provision of fuses or circuit breakers
  - of persons by the provision of:
    * insulation
    * 110-volt centre tapped to earth supply
    * residual current device (RCD) in 240-volt supply
    * means to discharge static electricity charges
- **reduced voltage**
  - use of a voltage below the supply level for ancillary services and to protect operators. Typically, reduced voltage levels are below 50 volts
- **residual current device (RCD)**
  - a device that senses leakage of current to earth and breaks the supply. Nominally senses leakage of 30 mA and breaks the circuit within 30 m secs.

Safe use of electricity **231**

17.2.1 Dangers from electricity

The following dangers are associated with electricity:

- shock:
  - usually from live wire to earth
  - causes muscular spasm
  - can interfere with the regular action of the heart and cause fibrillation of the heart muscles or complete cardiac arrest
  - can cause respiratory failure
- burns:
  - from contact with electric arc caused when conductors short circuit or equipment is overloaded
- arc eyes:
  - from ultraviolet rays when looking at electric arc or welding flash
  - symptoms like conjunctivitis
  - temporary condition lasting three or four days
  - does not affect contact lens
- fire:
  - from electrical arc
  - overloading of conductors
  - discharge of static electricity
- static:
  - caused when two materials are parted, e.g. web from roller; solvent being poured from container
  - high voltage, low current
  - causes spasm of voluntary muscles and violent body movement when injury results from hitting equipment not from the static itself.

17.2.2 Safe use techniques

This part is not concerned with the installation of electrical equipment – that should be done by trained and competent electricians – but with the use of electricity in the workplace.

- Faults:
  - if faults occur, do *not* interfere with electrical equipment – get a qualified electrician
- Circuit protection:
  - the function of a fuse or a circuit breaker is to prevent the downstream circuit from being subjected to current beyond its capacity
  - fuses allow some excess current to flow before they 'blow'
  - circuit breakers trip at the set current
  - all electrical circuits should be protected by a fuse or circuit breaker
  - fuses/circuit breakers should be rated to protect the equipment served
  - fuses/circuit breakers should not be uprated except by qualified electrician
- Isolating switches
  - every piece of equipment using electrical power should have its own isolating switch
  - isolating switches should have means for locking them in the OFF position
  - isolating switches should be labelled to identify equipment served
  - isolating switches must be within easy reach from the working floor level
- Maintenance
  - when working on equipment do not rely on the normal OFF switch – switch off at the isolator and lock it off
  - locking off should be by individual padlock

- use special multi-padlock calipers if more than one person is working on equipment
- strict rules should apply to locking-off procedures – see below
- Earthing
  - all electrical equipment, except where double insulated, must be earthed
  - earth circuits must be solid connected to earth and not pass through switches
- Access
  - access space at least 1m wide must be left in front of electrical switch-gear – a hand rail positioned 1 m out from switch-gear will keep space clear
- Protection:
  - new installations and existing equipment working in a hostile (wet) environment should have earth leakage protection such as residual current devices (RCDs)
  - all electrically driven machinery should have emergency stop switches which are:
    * readily identified
    * within easy reach of operators
  - interlock switches should be limit switches (*not* micro-switches)
  - actuation of single interlocking switches should be positive, i.e. safety circuit made when switch in the relaxed (non-operated) position
  - all conductors must be insulated – bare wires or exposed connections are *not* permitted
  - for portable equipment '110-volt centre tapped to earth' supply should be used
  - display placards:
    * summarizing the *Electricity at Work Regulations 1989*
    * showing means of artificial respiration (rescusitation).

17.2.3 Rules for locking-off

1. All electrical machinery should have isolators with locking-off facilities.
2. Before starting any maintenance, repair or other work that requires access into the machinery the isolator must be locked-off by padlock and identifying tag attached.
3. Each padlock should have only one key. There should be no duplicates or master keys.
4. Only the person who attached the padlock may remove it. Arrangements may need to be made to transfer padlocks (or keys) at shift change-overs.
5. If more than one person is working on the equipment, multi-padlock hasps should be used and each person attach their own padlock.
6. On major maintenance a single padlock can be used for a gang, in which case the supervisor/foreman carries the responsibility for the safety of the whole gang and for ensuring they are all clear of the equipment before removing the padlock.
7. Before removing the last padlock on completion of the work, the equipment must be checked to ensure all tools have been removed, guards replaced and the equipment is safe to operate.
8. Padlocks should either be issued on a personal basis or kept centrally and signed for at each use.
9. Loss of padlock keys should be reported to the supervisor and the written authority of a responsible manager obtained before the padlock is removed forcibly.

Safe use of electricity 233

10 In an emergency, if the 'owner' of a padlock is not available, the authority of a responsible manager should be obtained before a padlock is removed.
11 Any employee leaving a padlock on an isolator unnecessarily at the end of a shift should be brought back to work to remove it.
12 Breach of these rules should be subject to disciplinary action.

By following these various proven safe practices and rules full and safe benefit can be obtained from the use of electricity.

17.3 Safe use of portable electrical equipment

This section concerns any piece of electrical equipment that is carried about during its normal use and which uses a mains power supply. Typical equipment includes hand lamps, any hand-held power tools, portable pipe-threading machines, fans, etc.

17.3.1 Precautions

For various types of portable electrical equipment the following precautions should be taken:

- Supply:
  - 110-volt centre tapped to earth:
    * can be an installed supply or
    * from 240-volt supply via a suitable portable transformer.
  - if 240-volt supply a residual current device (RCD) must be used
  - in both cases the supply must include an earth connection
- Equipment:
  - be either:
    * earthed or
    * double-insulated
- Plug:
  - suitable for the supply outlet
  - properly wired including clamping of the cable sheath
  - the earth wire should have plenty of slack in the plug so it is the last wire to pull out of the terminals thus retaining earthing integrity
  - in good condition
  - fitted with the correct fuse
- Cable:
  - of suitable capacity for the appliance, both voltage and current
  - of flexible type
  - in good condition without any damage to the sheath
  - have earth conductor except in case of double-insulated equipment
  - inspected regularly
- Appliance:
  - in good condition and repair
  - properly wired
  - suitable for the supply voltage
  - the sheath of the supply cable securely clamped
  - properly earthed unless double-insulated
- Inspection:
  - all portable powered equipment (plug, cable and appliance) must be inspected regularly and have suitable dated tag attached.

17.3.2
Likely faults

The following faults can occur and are relatively easy to fix:

- damaged cable:
  - replace or insert proper connector at point of damage
- grommet missing at point of entry to appliance:
  - replace grommet
- sheath pulled out of clamp, i.e. the individual wires can be seen:
  - replace and clamp securely on the sheath
- broken plug:
  - replace.

These precautions are simple and easy to take but they do provide a high level of protection in the use of portable equipment. They are as applicable at home as they are at work.

18 Fire

After accidents, fire is one of the largest drains on industrial resources. Not only is it a drain but it is also a major cause of companies going out of business. A fire can remove overnight the manufacturing ability of a company and without that ability customers are lost. No customers, no business. So the prevention of fire can play an important role in the continuing viability of an organization.

The general powers of fire authorities to fight fires are contained in *The Fire Services Act 1947* which requires the fire brigade to operate in an efficient and organized manner, to ensure there is an adequate supply of water for fighting fires and gives them the right of entry to buildings if a fire is suspected.

The value of fire precautions at work has long been recognized and extensive requirements were contained in the *Factories Act 1961*. These requirements have been superseded by the *Fire Precautions Act 1971* (FPA) which has been supplemented by the *Fire Safety and Safety of Places of Sport Act 1987*. Although this latter Act was brought in largely as a reaction to the Hillsborough tragedy it also incorporates amendments to the FPA. New Regulations – the *Fire Safety (Workplace) Regulations 1997* – bring into UK law the fire precaution requirements of the Framework and Workplace Directives.

These two Acts and the new Regulations are considered in more detail in the following sections.

18.1 Fire legislation

This section summarizes the main requirements of the *Fire Precautions Act 1971* as amended by various legislation including the *Fire Safety and Safety of Places of Sport Act 1987*. The Act requires that premises that meet certain criteria must have a Fire Certificate. The premises involved include:

- factories
- offices
- shops
- railway premises
- hotels and boarding houses.

18.1.1 Responsibilities of employer/occupier

These are:

- to apply for a Fire Certificate
- to provide:
  - a fire-warning system
  - fire-fighting appliances (extinguishers, hoses, sprinklers, etc.)
  - adequate means of escape and keep it clear
  - fire training to employees including fire drills.

18.1.2 Fire certificate

This is required for premises that:

- provide sleeping accommodation:
  - for more than six persons or
  - above the ground floor or
  - below the ground floor
- are institutions providing treatment or care
- are places of entertainment, recreation or clubs
- are for teaching, training or research
- allow access to members of the public
- are used as a place of work where:
  - more than twenty people are employed on ground floor
  - more than ten people are employed other than on the ground floor, i.e. on the first floor or above or in a basement
  - explosives or highly flammable liquids in quantities above a level agreed with the Fire Authority are used or stored.

Exemptions can be granted at the discretion of the Fire Authority if the premises are:

- low risk and
- there are adequate fire precautions.

Low risk means:

- ground floor only
- ground plus first floor only
- ground plus first floor plus basement if there is half-hour fire barrier (both construction and doors) between basement and the ground floor.

Application for certificate is:

- to be on prescribed form, copy available from Fire Authority
- accompanied by a plan of the premises showing fire doors, walls, fire extinguishers, hose reels, fire alarm system, etc.
- to be made by the occupier or in multi-occupancy premises by the owner.

Premises are inspected by Fire Authority who will either:

- issue a notice detailing work to be done before a certificate is issued
- issue a Fire Certificate
- exempt the premises from need to have a certificate
- state that the premises do not need a Fire Certificate.

The Fire Certificate specifies:

- address of premises
- name of responsible person
- description and use of premises
- fire resisting walls, floors and doors
- means of escape:
  - escape routes
  - escape routes to be kept clear
  - fire doors
  - smoke doors

- fire fighting equipment:
  - extinguishers, hose reels, other means
  - their location
  - to be kept properly maintained
- fire warning arrangements
  - number of alarm points
  - position of alarm-warning devices
  - automatic fire-detection systems
- emergency lighting
- the number of employees allowed on the premises at any one time
- fire training for employees including:
  - evacuation procedure
  - use of extinguishers
  - fire drills
- maximum quantities of explosives and highly flammable materials that may be kept or used on the premises
- other precautions considered necessary for the premises
  - Fire Authority can make a charge for issuing a Fire Certificate.

Once the Fire Certificate has been issued to the occupier it is to be kept on the premises. Any proposed building alterations that affect the certificate must be notified to Fire Authority before work commences.

18.1.3 Enforcement

- by Fire Authority (local Fire Brigade) through Fire Prevention Officers
- except where there is a high fire risk from materials being processed when by HSE Inspector
- Inspectors must carry an official warrant of authority.

18.1.4 Powers of Inspectors

- right of entry into premises relating to the issue of or exemption from a Fire Certificate
- if premises used as a dwelling, twenty-four hours notice to be given
- to make enquiries and obtain information relevant to an inspection
- to inspect the Fire Certificate
- to issue a 'Steps to be Taken' Notice or an Improvement Notice when fire safety measures not up to standard
- to issue a Prohibition Notice when fire risk considered serious
- must not give information obtained from an inspection to a third party.

18.1.5 Appeals

Appeals against any decision of the Fire Authority are taken to a Magistrate's Court.

18.1.6 Legal action

For a breach:

- triable:
  - summarily in a Magistrate's Court
  - on indictment in the Crown Court
- defence that accused *took all reasonable precautions and exercised all due diligence* is allowed.

Although Fire Officers have been given similar inspection powers to HSE inspectors, they are not authorized to present cases in court.

238 Fire

18.1.7 1997 Regulations

The Fire Precautions (Workplace) Regulations 1997 add little to the existing requirements so that where appropriate fire safety measures such as:

- fire certificate
- means of escape
- fire fighting equipment
- fire training for employees

are in place little further action is required.

However, these new Regulations do require employers to:

- carry out an assessment of the risk of fire
- ensure there are means for the early detection of fire
- ensure there are adequate means of escape
- provide suitable fire fighting equipment and keep it readily available.

18.2 Causes of fire and precautions

Fires don't just happen, they are caused when three elements are present. These are:

1 Oxygen

- normally from the air which contains 20% oxygen
- can be given off by oxidizing chemicals such as nitrate fertilizers.

2 Fuel

- can be anything combustible:
  - with solids the smaller the pieces the more easily it ignites
  - with liquids, the lower its flash point the more readily it ignites
  - with gases the concentration needs to be within its flammability limits.

3 Ignition

- caused by any source that will raise the temperature above the flash or ignition point, including:
  - cigarette ends
  - electrical sparks and short circuits
  - static electricity
  - heated equipment and over heated bearings
  - heating pipes
  - sparks from welding and burning operations.

All three need to be present for a fire to start. Remove any one element and the fire dies.

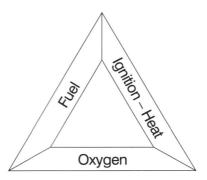

Figure 18.1 The fire triangle

Fire

Causes of fires	Precautions
Bad housekeeping	– regular cleaning with rubbish disposed of into suitable receptacles
Electrical	– do not overload circuits, ensure fuses/circuit breakers operate at correct current level – provide adequate cooling air flow – do not allow fluff to build up on electrical equipment – ensure electrical contacts are kept clean
Hot bearings	– keep clean and well maintained
Heating pipes	– ensure all combustible material is stored well clear of pipes
Welding and gas cutting	– remove combustible material from area – have suitable extinguishers positioned nearby – post fireman during actual work and for half-hour afterwards
Smoking	– prohibit or provide special smoking areas/bays with ash trays and fire extinguishers
Oils and solvents	– store outside – use only enough for day or shift – use non-spill containers for handling – use earth link when pouring
Oily rags	– dispose of in non-flammable (metallic) bins.

18.2.1 Heat transfer

Heat transfer can be by:

1. **Conduction** – the carrying of heat along or through materials
2. **Convection** – heat carried by rising currents of hot air
3. **Radiation** – the emission of long wavelength infra-red rays.

18.2.2 Fire spread

Fires can spread easily:

- across open plan layout:
  - normally spreads at high level therefore provide high-level ventilators/extractors to draw flames out of building
  - divide up area into fire compartments
- through gaps or holes in fire-break barriers
  - ensure all gaps in fire-break barriers are plugged with fire resisting material particularly round pipe and cable entries
  - ensure fire-break doors are kept shut or have means for closing them automatically if fire occurs
- along channels and gullies in which flammable liquids and heavier-than-air gases can flow:
  - for liquid spills:
    * bund round container
    * use suitable absorbent granules
  - for heavier-than-air gases and fumes ensure the area is well ventilated at low level

- across the surface of dusts in roof spaces
  - keep roof spaces cleaned
  - build fire breaks in roof space
- carried through ventilating and exhaust ducting
  - switch off ventilating fans
  - provide fire extinguishant in the ducts
- along corridors
  - fit smoke/fire doors and keep them shut or fit magnetic releases linked to fire-alarm system
- in lift shafts
  - shaft well should be enclosed
  - fit smoke/fire doors on landings at access to work areas and corridors.

The main thrust of fire precautions is to prevent a fire starting in the first place, but if a fire does occur the next priority is to ensure the layout of the building is such that it will stop it spreading and hence keep damage and disruption to a minimum.

18.3 Fire-fighting and extinguishers

This refers to 'first-aid' fire-fighting and not the fighting of major conflagrations.

Most fires start out very small and can be safely tackled at that stage by means of hand-held extinguishers. However, at no stage should employees put themselves at risk in attemping to put a fire out. They should always place themselves between the fire and the escape route.

18.3.1 Fire-fighting equipment

The types and number of equipment are determined by:

- requirements of the Fire Certificate
- advice of the local Fire Prevention Officer (for which you may have to pay)
- advice from fire insurer's surveyor
- pressure from a fire extinguisher salesman (treat their recommendations on numbers and the need for replacement units with suspicion – always check recommendations with the Fire Prevention Officer before buying).

18.3.2 Types of equipment

There are two types of equipment: installed and hand-held.

1 Installed equipment

- hose reels:
  - need reliable water supply
  - suitable number to cover whole work area
  - can be a nuisance if stop cock leaks
  - can be arranged to turn water on as hose is pulled out
- sprinklers:
  - usually in high-risk areas
  - liked by fire insurers who reduce premiums
  - sprinklers spray covers approx. 10 m^2 (100 ft^2)
  - will contain the fire but can cause a great deal of damage to stock
  - need guaranteed supply of water at pressure
  - expensive to install and maintain
- halogen gas:
  - in computer and electrical control gear
  - suppresses fire but does not remove heat so fire can restart when gas turned off or runs out
  - being replaced by water in some computer installations

- carbon dioxide:
  - in electrical substations
  - for solvents on printing machines
  - is an asphyxiant so all employees must be clear before gas turned on.

2 Hand-held extinguishers

There are five basic types of these, each colour-coded to identify its contents and the sort of fire for which it is suitable.

Type	Colour	Suitable for
Water	red	carbonaceous materials – wood, paper, coal, etc.
Foam, including aqueous film forming foam (AFFF)	cream	carbonaceous materials, liquids whether soluble in water or not
Carbon dioxide	black	liquids whether soluble in water or not; fires in electrical equipment
Dry powder	blue	liquids whether soluble in water or not; fires in electrical equipment
Vapourizing liquid	green	liquids whether soluble in water or not; fires in electrical equipment

EU Directive requires that all extinguishers supplied after May 1997 must be painted red but allows 5% of body to be a different colour. Existing extinguishers need not be repainted.

18.3.3 Classes of fires

Class	Materials	Extinguishant
A	Organic materials – wood, coal, paper	Water, AFFF
B(i)	Liquids and liquefiable solids soluble in water – acetone	Foam, vapourizing liquids, CO_2, dry powder, AFFF
B(ii)	Liquids and liquefiable solids not soluble in water – petrol, fats, wax	Foam, AFFF, vapourizing liquids, CO_2, dry powder
C	Gases and liquefied gases – propane, butane	Isolate supply, dry powder
D	Metals – magnesium, aluminium	Special dry powder

Electricity can cause a fire in any class – do not use water, foam or AFFF extinguishers

242 Fire

Fire-fighting equipment should be:

- mounted on red-painted fire points
- well identified
- located adjacent to fire exit
- can include an alarm point.

Training to be provided to:

- all staff in fire drill
  - what to do when alarm sounds
  - assembly points
  - roll call
  - role of fire marshals
- supervisors and selected staff:
  - in use of extinguishers
- volunteer fire teams:
  - in first-aid fire-fighting.

Action on finding a fire:

1. Sound the alarm (break glass).
2. Call fire brigade.
3. Use extinguisher if safe to do so.
4. Evacuate premises by quickest route.
5. Go to assembly point for roll call.

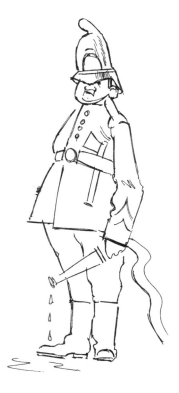

Where's the fire then?

Like oak trees, fires start from small beginnings. If they can be tackled at that stage an enormous amount of disruption and loss can be prevented. However, in no circumstances should employees put themselves at risk in tackling even a small outbreak.

18.4 Safe use of flammable substances

Many highly flammable substances are used at work and in spite of their flammable nature can be used quite safely if certain precautions are taken.

Liquefied petroleum gases (LPG)

- butane and propane
  - stored in bulk or in portable cylinders
  - cylinders to be:
    * stored out of direct sunlight
    * kept in well ventilated area
    * fitted with pressure relief valve
  - after use turn off at valve on cylinder
  - do not rely on the integrity of hose to prevent leakage
  - check hose regularly for signs of cracking and splitting
  - use in well ventilated areas
  - no smoking in area where LPG in use or store.

Welding and burning gases

- oxygen, hydrogen,
  - stored in high pressure cylinders
- acetylene
  - stored by dissolving in absorbent medium in high pressure cylinder
- all flammable gases
  - store cylinders of gases separately, i.e. do not mix cylinders of different gases in storage
  - store cylinders out of direct sunlight
  - store cylinders upright
  - in use restrain cylinders upright in proper frame
  - use anti-flashback fitting in gas line
  - check gas hoses regularly for leaks or damage.

Flammable liquids

- petrol, paraffin, white spirit, etc.
  - store outside building
  - bund the storage area
  - use earth link when transferring liquids
  - use drip tray in decanting area
  - transport in non-spill containers
  - use only in well ventilated areas
  - ban smoking in area
  - operators to wear protective gloves.

Oxygen enrichment

- occurs when using oxgen in confined or restricted spaces:
  - ensure space is well ventilated
  - turn oxygen supply off at cylinder or manifold valve when not being used – even for short tea or meal breaks.

Foam plastics

- store bulk away from work area
- have only supplies for day or shift in work area
- enforce no-smoking
- keep escape routes clear of material
- dispose of scraps in suitable containers.

19 Environment

A good working environment has a great influence on the health of the employees and on their attitude towards their work. Also, the atmosphere within the workplace and how it gets rid of harmful fumes has a large effect on the local community. *The Control of Substances Hazardous to Health Regulations 1994* (COSHH) were aimed at providing protection for the health of employees inside the workplace and the *Environmental Protection Act 1990* (EPA) set out to protect the community from a whole range of potential health hazards including those caused by work activities. The method of disposal of hazardous wastes is regulated by *The Special Waste Regulations 1996*.

COSHH is dealt with in Section 11.1. EPA and Special Waste Regulations are dealt with in Section 19.1.

19.1 The Environment Protection Act 1990

This Act set out to deal with the whole question of pollution of the environment and is aimed at regulating pollution from industrial, domestic and community sources. This section is concerned only with pollution resulting from work activities.

The Act:

Defines:

- **pollutants** as:
  - solid wastes for discharge onto land
  - liquid wastes whether discharged on to land or into waterways
  - discharges into the atmosphere
  - noise in the community

- **controlled waste** as:
  - household waste
  - industrial waste
  - commercial waste

- **special waste** as:
  - controlled waste that is so dangerous that it requires special disposal procedures, i.e. dangerous to life or liquids with a flash point of 21°C or less.

Has a three-pronged approach:

1. Air Pollution Control (APC):
   - deals with discharges into the atmosphere
   - enforced by local authority officers
2. Integrated Pollution Control (IPC):
   - deals with discharges to:
     * the air
     * the ground
     * waterways
   - enforced by Environment Agency Inspectors
3. Control of discharges on to land.

Gives Minister powers to:

- specify control standards to be met
- 'prescribe' processes requiring authorization
- issue authorization for prescribed processes
- place conditions on authorizations requiring the use of *best available techniques not entailing excessive cost* (BATNEEC) (not to be confused with CATNAP – cheapest available techniques not attracting prosecution)
- charge for issuing an authorization.

Appeals concerning the issue, revocation or alteration of authorizations heard by the Secretary of State.

No one allowed to carry on a 'prescribed process' without due authorization.

Enforcement by inspectors of:

- the Environment Agency (EA) appointed by the Minister
- river purification authority
- local authorities.

Powers of inspectors:

- entry into premises where prescribed processes are carried out
- issue *Enforcement Notices* (similar to Improvement Notices under HSW)
- issue *Prohibition Notices* (again similar to Prohibition Notices under HSW)
- carry out investigations
- measure and take photographs
- take samples
- require person to give information
- examine records
- obtain assistance with investigation
- if imminent danger, to render substances harmless.

Enforcing authorities to:

- keep register of prescribed processes and authorizations
- make it available for inspection by members of the public.

Appeal against a Notice is to the Secretary of State.

> In a prosecution, a defence for accused is to prove he used BATNEEC.
>
> A *waste management licence* is required for keeping, treating or disposing of controlled waste on land.
>
> Local authority officers inspect for *statutory nuisances* including:
>
> - discharges into the atmosphere
> - noise emitted from a premises that is prejudicial to health or a nuisance (see Sections 12.3 and 12.4).

Maintaining a good environment at work is sound commercial sense and has many benefits for the workforce and for the local community.

19.2 Safe and healthy working environment

A large portion of a person's life is spent in the work environment. A good environment will ensure they stay healthy in both body and mind and be able to enjoy a high quality of life. This section considers some of the factors in the work environment.

- Atmosphere:
  - workplace must be provided with sufficient quantity of fresh or purified air
  - should be free from contaminants such as dusts and fumes [see Section 11.6]
  - extract dust and fumes at source and filter before discharging outside the building
  - have good natural ventilation wherever possible
  - openable windows
  - no-smoking rule in work area
  - provide special smoking rooms or bays if necessary
  - if forced ventilation (air conditioning):
    * ensure no draughts from outlets
    * check noise levels
    * provide local controls
    * check system for legionella
  - with free-standing fans:
    * ensure blades are securely guarded
    * ensure they do not cause draughts
    * protect trailing cables.
- Lighting:
  - to be bright enough to enable work to be done without strain to eyes
  - walkways adequately lit
  - local lighting for precise work
  - good overall general lighting
  - no points of light likely to cause glare
  - use natural light wherever possible
  - provide blinds or shade from bright sunlight.
  - see CIBSE 'Code for Interior Lighting'
- Cleanliness:
  - work area should be cleaned regularly
  - rubbish disposed of in proper receptacles.

- Overcrowding:
  - ensure each person has at least 11m^3 (400 cu ft)
  - allow for space taken up by large pieces of equipment
  - provide adequate gangways between work stations.
- Temperature:
  - is required to be *reasonable*
  - not specified but normally taken to be:
    for sedentary work a minimum of 16°C (60.8°F)
    for hard physical work a minimum of 13°C (55.4°F)
  - a suitable number of thermometers should be positioned about the workplace.
- Noise:
  - should not be excessive:
    * in manufacturing areas less than 85 dB(A) (see Section 12.1)
    * in offices, laboratories, libraries, etc. not in excess of 40 dB(A).

A clean and healthy working environment is good business sense reducing likely illnesses (and corresponding absences) to a minimum and providing an atmosphere that encourages the workpeople to give of their best.

Appendix 1 Abbreviations

Abbreviation	Meaning
ACAS	Advisory, Conciliation and Arbitration Service
ACOP	Approved Code of Practice
ACTS	Advisory Committee on Toxic Substances
ADR	European Agreement concerning the International Carriage of Dangerous Goods by Road
APC	Air pollution control
BATNEEC	Best available technique not entailing excessive costs
BSC	British Safety Council
CAS number	The number assigned to a substance by the Chemical Abstract Service
CDM	The Construction (Design and Management) Regulations 1994
CE	The mark stamped on machinery and equipment to show that it conforms with EU directives
CEN	Comité Européen de Normalization – European Committee for Standardization (for mechanical equipment)
CENELEC	Comité Européen de Normalization Electrotechnique – European Committee for Electrical Standardization
CHASE	Complete Health and Safety Evaluation
CHIP	The Chemical (Hazard Information and Packaging for Supply) Regulations 1994 & 1996
CIBSE	Chartered Institution of Building Services Engineers
CITB	Construction Industry Training Board
COP	Code of Practice
COSHH	The Control of Substances Hazardous to Health Regulations 1994
CS	Chemical series (of Guidance Notes)
dB(A)	'A' weighted decibel
DoE	Department of the Environment
DSE	Display screen equipment
DTI	Department of Trade and Industry

EcoSoC	Economic and Social Committee of the European Union
EEC	European Economic Community (now European Union)
EFTA	European Free Trade Association
EH	Environmental health series (of Guidance Notes)
EHO	Environmental Health Officer
EINECS	European Inventory of Existing Commercial Substances
EL	Employer's Liability (insurance)
ELINCS	European List of Notified Chemical Substances
EN	European Normalization (prefix letters to harmonized European standards)
EP	European Parliament
EPA	The Environmental Protection Act 1990
ESR	Essential safety requirements
EU	European Union
FA	Factories Act 1961
FMEA	Failure mode and effect analysis
FPA	Fire Precautions Act 1971
FTA	Fault tree analysis
GS	General Series (of Guidance Notes)
HAVS	Hand and arm vibration syndrome
HAZAN	Hazard analysis
HAZOPS	Hazard and operability studies
HMSO	Her Majesty's Stationery Office
HS(G)	Health and Safety Guidance booklets
HS(R)	Health and Safety Regulation booklets
HSC	Health and Safety Commission
HSE	Health and Safety Executive
HSW	The Health and Safety at Work, etc. Act 1974
Hz	Hertz (cycles per second)
IAC	Industry Advisory Committee
IEC	International Electrotechnical Commission (for International electrical standards)
IEE	Institution of Electrical Engineers
IOSH	Institution of Occupational Safety and Health
IPC	Integrated pollution control
IP rating	Index of Protection (an international rating system – see BS 5420 & 5490)
ISO	International Standards Organization (for International mechanical standards)
ISRS	International Safety Rating System
IT	Information technology
L	Legal Series of booklets
LEV	Local exhaust ventilation
LPG	Liquified petroleum gas
MEL	Maximum exposure limit
MHOR	The Manual Handling Operations Regulations 1992

Appendix 1 **251**

MHSW	The Management of Health and Safety at Work Regulations
MIIRSM	Member of the International Institute of Risk and Safety Managers
MS	Medical Series (of Guidance Notes)
NEBOSH	National Examination Board in Occupational Safety and Health
NVQ	National Vocational Qualification
OES	Occupational exposure standard
OSRP	The Offices, Shops and Railway Premises Act 1963
PL	Public Liability (insurance)
PM	Plant and Machinery Series (of Guidance Notes)
PPE	Personal protective equipment
PUWER	The Provision and Use of Work Equipment Regulations 1992
RCD	Residual current device
RID	European Agreement concerning the International Carriage of Dangerous Goods by Rail
RoSPA	The Royal Society for the Prevention of Accidents
RPE	Respiratory protection equipment
RSI	Repetitive strain injury
RSP	Registered Safety Practitioner (of IOSH)
SME	Small and medium enterprise
SMSR	The Supply of Machinery (Safety) Regulations 1992
SQWG	Social Questions Working Group (of the EU)
SVQ	Scottish Vocational Qualification
TREMCARD	Transport emergency card
TWA	Time weighted average
UK	United Kingdom
UN	United Nations
VCM	Vinyl chloride monomer
VWF	Vibration white finger
WATCH	Working Group on the Assessment of Toxic Chemicals
WHSWR	The Workplace (Health, Safety and Welfare) Regulations 1992
WRULD	Work related upper limb disorder

Appendix 2 List of Statutes

Boiler Explosions Act 1881	*20*
Building Act 1984	*22*
Carriage of Dangerous Goods (Classification, Packaging and Labelling) and Use of Transportable Pressure Receptacles Regulations 1996	*127; 137*
Chemical (Hazard Information and Packaging for Supply) Regulations 1994/1996	*127; 133*
Construction (Design and Management) Regulations 1994	*188; 192; 203*
Construction (Head Protection) Regulations 1989	*119*
Construction (General Provisions) Regulations 1961	*191*
Construction (Health, Safety and Welfare) Regulations 1996	*188; 191; 200; 203*
Construction (Health and Welfare) Regulations 1966	*191*
Construction (Working Places) Regulations 1966	*191*
Consumer Protection Act 1987	*30*
Control of Asbestos at Work Regulations 1987	*43; 119*
Control of Lead at Work Regulations 1980	*43; 119*
Control of Pollution (Special Waste) Regulation 1980	*155; 245*
Control of Substances Hazardous to Health Regulations 1994	*43; 114; 127; 245*
Dangerous Machines (Training of Young Persons) Order 1954	*58*
Disability Discrimination Act 1995	*30; 63*
Disabled Persons (Employment) Act 1944	*62*
Electricity at Work Regulations 1989	*226; 230; 232*
Employment Protection Act 1975	*62*
Employer's Liability (Compulsory Insurance) Act 1969	*30; 66*
Environmental Protection Act 1990	*29; 160; 245*
Equal Pay Act 1970	*62*
European Communities Act 1972	*18; 21*
Factories Act 1961	*18; 20; 21; 187; 215; 235*
Factory and Workshop Act 1878	*20*
Fire Precautions Act 1971	*235*
Fire Safety and Safety of Places of Sport Act 1987	*235*

Fire Safety (Workplace) Regulations 1997	235; 238
Fire Services Act 1947	235
Health and Safety at Work, etc. Act 1974	18; 20; 21; 34
Health and Safety (Consultation with Employees) Regulations 1996	59
Health and Safety (Display Screen Equipment) Regulations 1992	43; 122
Health and Safety (First Aid) Regulations 1981	117
Health and Safety (Safety Signs and Signals) Regulations 1996	79; 82
Health and Safety (Young Persons) Regulations 1997	56
Hoists Exemption Order 1962	216
Industrial Tribunals (Improvement and Prohibition Notices Appeals) Regulations 1974	7
Ionizing Radiations Regulations 1985	115 ; 119
Lift Regulations 1997	186; 214; 216
Management of Health and Safety at Work Regulations 1992	25; 34; 43; 53; 87
Manual Handling Operations Regulations 1992	43; 209
National Insurance (Industrial Injuries) Act 1946	30; 66
Noise at Work Regulations 1989	119; 158
Occupier's Liability Act 1984	30
Offices, Shops and Railway Premises Act 1963	18; 20; 21
Offshore Safety Act 1992	30
Personal Protective Equipment at Work Regulations 1992	43; 119
Pressure Systems and Transportable Gas Containers Regulations 1989	183
Provision and Use of Work Equipment Regulations 1992	43; 167; 171; 214; 226
Race Relations Act 1976	29; 62
Reporting of Injuries, Diseases and Dangerous Occurrences Regulations 1995	99
Road Traffic Regulations Act 1984	82
Road Vehicles (Construction and Use) Regulations 1986	225
Safety Representatives and Safety Committees Regulations 1977	59
Sex Discrimination Acts 1975 & 1986	29; 62
Single European Act 1986	15; 20
Social Security Act 1989	66
Social Security (Industrial Diseases) (Prescribed Diseases) Regulations 1980	66
Supply of Machinery (Safety) Regulations 1992	17; 43; 167; 173; 187; 214
Trade Union Reform and Employment Rights Act 1993	29; 63
Workmen's Compensation Act 1897, 1906 & 1925	20; 65
Workplace (Health, Safety and Welfare) Regulations 1992	34; 69; 79; 191

Index

Absorption of poisons 110
ACAS (Advisory Conciliation and
  Arbitration Service) 62
Access:
  equipment 200
  space 232
  towers 202
  ways 74
Accident: 95
  cause of 97
  definition 94
  investigation 97
  reporting 99
  triangle 96
Accident prevention:
  principles 94
  techniques 95
  role of engineering 97
Acetylene 243
Acts 14, 20, 88
  Single European 15
ACTS (Advisory Committee on Toxic
  Substances) 149
AIDS 113
Air pollution control 246
Alpha (α) particles 114
Anthrax 113
Appointed person 118
Approved bodies 169, 186
Approved Carriage List 137, 142, 143
Approved person 139
Approved Supply List 128, 133, 142
Arc eye 231
Asbestos 197
Asthma, bronchial 112
Asphyxia 112
Atmosphere, working 247
Audiometer 162
Authority:
  Fire 23, 25
  Local 23
'A' weighting 163

Banksman 85, 220, 225
BATNEEC 246
Bench, Queen's 10
Beta (β) particles 114
Biological agent 128
Body, functions of 105
Breach of statutory duty 66
Bremsstrahlung 114
British Safety Council 48, 91
British standard 179
  BS 5304 176
Bronchial asthma 112
BSI (British Standards Institution) 17, 48, 89
Building, safe maintenance 180
Butane 243
Bylaw 21

Cabinets, filing 79
Carcinogens 112, 128
Carpal tunnel syndrome 113
CAS number 142
Category of danger: 133
  health effects 141
  physio-chemical properties 141
CDM (Construction (Design and
  Management) Regulations 1994) 188, 203
CE mark 167, 169, 170, 217
CEN (Comité Européen de Normalization) 17
CENELEC (Comité Européen de
  Normalization Electrotechnique) 17
Certificate of Adequacy 169, 170
Charge, electrical 229
Chartered Institution of Building Services
  Engineers 71
Chemicals:
  control measures 152
  for supply 133
  for transport by road and rail 133
  handling 152
  labels 135

Chemicals – *continued*
  packaging 138
  plants, safe maintenance 182
  preventitive measures 151
Children 27, 55
CHIP (Chemical (Hazard Information and Packaging for Supply) Regulations 1994) 127
Circuit breakers 231
Circuit protection, electrical 231
Classification of chemical dangers 141
Cleanliness 247
Client 189
Clothes, storage 78
Code of Practice, Approved 21, 89
Coffer dams 193
Compensation 65
Competence 229
Competent:
  person 35, 159, 195, 200, 220
  supervision 35, 56
Complete Health and Safety Evaluation 48
Conduction, of heat 239
Confined spaces 70, 179, 181
Conformity assessment procedure 172, 217
Construction 188
Consultation, joint 59
Contract workers 25
Contractors: 27, 198
  main 192
  principle 189, 191
  responsibilites 199
  sub- 191
Controlled waste 245
Convection 239
Conveyors 214, 221
Corrosives 111
COSHH (Control of Substances Hazardous to Health Regulations 1994) 92, 114, 127, 245
Costs, of failure 97
Council of Ministers 1
Court:
  Chancery 10
  Civil 6
  County 7
  Criminal 6
  Crown 5
  Divisional 10
  European, of Justice 10
  Family 10
  High 10
  Magistrate's 5, 6, 23
  of Appeal 7
Cranes 187, 214, 219
  mobile 198

Damages, claims for 65, 67
Danger 39
  indication of 134
  signs 140
Data:
  retrieval 92
  storage 92
  storage systems 93

Date, of knowledge 67
Decibel 163
Decided cases 6
Decision, EU 15
Declaration of Conformity 168, 170, 217
Declaration of Incorporation 168, 170
Demolition 193, 203
Designer 189
Developer 189
Direct entry of poisons 110
Directive, EU 15, 21, 167, 171
Disabled persons 62
Discharges, control of 246
Discipline 62
Discrimination:
  race 7, 29, 62
  sex 7, 29, 62
Diseases, of metals 112
Document, consultative 14
Domino theory (Heinrich's) 94
Doors 5
  fire 76
  smoke 76
Dosimeter 163
Double insulation (electrical) 230, 233
DSE (Display screen equipment) 81, 122
Due diligence 136, 226, 237
Dusts 111, 150
Duties 36
  of care 66
Duty holder 230

Ear 161
  muffs 159, 166
  plugs 159, 166
  protection zone 159
Earthing 230, 232, 233
Eating facilities 79
EC number (of chemical) 134
EC type examination certificates 170
EFTA (European Free Trade Association) 17
EHO (Environmental Health Officer) 24
EINECS (European Inventory of Existing Chemical Substances) 134
Electrical:
  equipment: 80
  portable 233
  protection 230, 232
Electricity 226
  safe use 229
  static 231
ELINCS (European List of Notified Chemical Substances) 134
EMAS (Employment Medical Advisory Service) 22
Emergency:
  escape routes 194
  plans 194
  stop 175, 228
Enforcement 22, 246
  Notice 246
English Justice 1
English law, sources of 6
English legal system 3
English system of justice 3

Environment 29, 123, 246
  checks 45
  working 247
Environmental Agency 246
Escalators 76, 216
ESR (essential safety requirement) 168, 170, 217, 218
European Economic Community (EEC) 1, 15
European Union (EU) 1, 3, 15
EU legislation:
  decisions 15
  directives 15
  regulations 15
Examination:
  scheme of 184
  periodic 220
Excavations 193, 196, 206
Excess current protection 230
Explosives 193, 205
Exposure limits 149
Eye 107

Falling objects 74, 196
Falls, from height 73, 192, 196
Fencing 178
Film badges 114
Fines 11
Fire 235
  causes 239
  certificate 93, 235
  classes of 241
  escape routes 81
  exit doors 81
  extinguishers 81, 235, 240
  fighting equipment 237, 240
  means of escape 235
  precautions 80
  Prevention Officers 25
  spread 239
  triangle 238
  warning system 235, 237
First aid 117
  boxes 118
First instance 7
  European Court of 15
Flammable liquids 243
Floors 72, 79
FMEA (Failure mode and effect analysis) 43
Foam plastics 244
Fragile roofs 192
FTA (Fault tree analysis) 43
Fumes, offensive/noxious 35, 80, 150
Fumigation 132
Furniture 81
Fuses 231

Gamma (γ) rays 114
Gassing 112
Gas welding 243
Gates 75
Geiger–Muller tubes 114
Glazing 74
Grab sampling 130
Guards 174, 177

Guard:
  material 178
  rails 74

Handling:
  manual 209
  mechanical 214
HAVS (hand and arm vibration syndrome) 113
HAZAN (Hazard analysis) 43
HAZCHEM:
  code 153
  sign 155
HAZOPS (Hazard and operability study) 43
Hazard 176
  definition 39, 94
  elimination 95
  warning board for vehicles 154
  identification 40, 44, 96
  rating 41
  reduction 40, 96
  warning signs 154
Health:
  hazards, causes of 111, 204
  surveillance 26, 131
Health and safety file 190
Health and safety plan 189
Hearing:
  acuity 162
  protection 159
  shift of threshold of 161
  threshold of 162
Heat 113
Heinrich's domino theory 94
Hepatitis 113
Her Majesty' Stationery Office (HMSO) 4
Hertz 164
Hoists 187, 214
Hold-on control 179
Housekeeping 71, 197
HSC (Health and Safety Commission) 18, 22
HSE (Health and Safety Executive) 18, 22
Human factors 63
Humidity 113

IAC (Industry Advisory Committee) 25, 48
Industrial Tribunal 4
IEC (International Electrotechnical Commission) 17
IEE Wiring Regulations 226
Industrial relations 61
Information: 27, 36, 60, 88, 98, 123, 131, 173, 210
  sources of 88
  retrieval 93
Ingestion 109
Inhalation 109
Injuries, to be reported 99
Integrated pollution control 246
Ionizing:
  chambers 114
  radiations 113, 114

Inspectors: 88, 90
  powers of 23, 235, 246
Inspections: 195
  matters for 45
Insulation, electrical 230
Insurance:
  employer's liability 11, 12, 67, 101
  fire 68
  in health and safety 65
  plant and machinery 68
  product liability 68, 190
  public liability 67, 221
Interlocutory hearing 12
IOSH (Institution of Occupational Safety and Health) 48
IP rating 230
Irritants 112
Isolation, electrical 230
ISO (International Standards Organization) 17
International Safety Rating System 48

Judge:
  Circuit 7
  District 7
Justice, summary 6

KISS principle 92
Knowledge, date of 67

Labelling of chemicals 133
Labels, information to be contained 135
Ladders 73, 192
Law:
  Civil 5
  Common 5
  Criminal 5
  Equity 5
  Private 5
  Public 5
  Rule of 6
    Statute 4
    subordinate 14
Legionella 113
Legislative harmonization 167
Leptospirosis 113
LEV (local exhaust ventilation) 130
Lifting:
  accessories 186, 219
  equipment 171, 186, 214
  machines 187, 219
  mechanics 210
  tackle 187, 214, 219
Lifts 187, 214
  'Bean-stalk' type 202
  Paternoster type 215
  scissor 216
  'Snorkel' type 202
Lighting 71, 82, 176, 195, 247
  emergency 237
Limited inch 179
Liquified petroleum gas 243

Litigation 5
Locking-off 50, 175, 231
  rules for 232
Lords, House of 7, 14

Machinery 168
  safe maintenance 181
Magistrate, Stipendiary 6, 10
Maintenance 179
  precautions 180
  safety in:
    buildings 180
    chemical plants 182
    confined spaces 181
    machinery 181
Management, techniques for accident prevention 96
Manual handling 209
Manufacturers 91
Maternity rights 63
Means of escape in case of fire 235
Medical examinations 131
MEL (maximum exposure limit) 128, 249
Method statement 190
Micro-organisms 113
Mobile machinery 171, 198
Mothers, nursing 26, 28
Multi-occupancy 27

National Radiation Protection Board 115
NEBOSH (National Examination Board in Occupational Safety and Health) 48
Negligence 6, 66
  contributory 67
Near miss 96
  definition 94
Neutrons 114
Noise 113, 158, 248
  abatement notice 160
  assessment 159
  'balancing' 166
  community 160
  control techniques 164
  daily personal exposure 158
  first action level 158
  induced hearing loss 161
  levels 162
  measurement 162
  meter 162
  peak action level 159
  second action level 159
Notice:
  Enforcement 246
  Improvement 7, 23
  Prohibition 7, 11, 23, 246
Notified body 217
Nuisance 6
  statutory 160, 247

Occupier 199
Octave band centre frequency 164
OES (occupational exposure standard) 128, 149
Order (statutory) 14, 21, 88

Organization:
  employee's 90
  employer's 90
  techniques 46
  types 46
Overcrowding 248
Oxygen enrichment 244
Ozone 80

Paper:
  Green 13
  White 13
Parliament 5
  Supremacy of 6
Permit-to-work 50, 51
Person in control 69
Photocopiers 80
Planning supervisor 189
Platform 73
  working 192
Poisons 111
Policy, safety 35, 36
Pollutant 245
Pollution:
  control of air 246
  integrated control 246
Portable electrical equipment 233
Power lines, overhead 197
PPE (Personal protective equipment) 35, 119
Prescribed process 246
Pressure systems 183
Principle contractor 189, 191
Printing machines 80
Probability 39, 41
Propane 243
Prosecution 11
Protective clothing 175
Protective techniques for machinery 174

Qualified majority voting 15

Radiations:
  dose rate 114
  exposure levels 114
  heat 239
  ionizing 113, 114
  non-ionizing 116
  Protection Adviser 115
  Protection Supervisor 115
Reduced voltage 230
110V centre tapped to earth 232, 233
Regulation: 14, 20, 88
  European 15
Residual current device 230, 232, 233
Relevant liquid 183
Repetitive strain injury 213
Report:
  accident 45, 92, 99
  near miss 45
  Robens 21
  writing 91

Res ipsa loquitur 67
Respiratory:
  protective equipment 129
  system 107
Responsibilities 34, 88, 199
Right of civil action 136
Risk:
  assessment 26, 38, 49, 56, 82, 94, 96, 129, 209, 238
  assessment strategy 39
  assessor 39
  definition 39, 94
  design assessment 43
  extent of 39
  number 142
  phrase 134, 142, 144
  rating 41
  residual 40, 96
Road vehicles 225
Roles, in organization 47
Rome, Treaty of 15
Rooms:
  changing 78
  rest 78
RoSPA (Royal Society for the Prevention of Accidents) 48, 91
Routes:
  access 200
  of entry 109
  pedestrian 75
  traffic 75, 193
  vehicular 75
Royal Assent 14
RSI (repetitive strain injury) 105, 113

Safety:
  advice 87
  adviser 26, 47, 87, 90, 98
  arrangements 38
  audit 44
  committee 35, 47, 59, 61
  consultant 87, 91
  culture (tone) 33
  data sheets 134
  hooks 220
  number 142
  office 79
  organization 37, 46, 48
  performance 33, 61
  phrase 134, 142, 147
  policy 35, 36, 37, 190
  representative 35, 47, 54, 59, 60, 90, 98
  sampling 44
  signals 82, 85
  signs: 82
    colours 83
    shapes 83
  survey 44
  tours 44
Safe systems of work 38, 49, 174, 178
Scaffolding 73, 197, 200
Seating 72
Sensitization 112
Severity of risk 41

Shelves 80
Signs:
    fire fighting 84
    information 85
    mandatory 84
    prohibition 83
    warning 84
Signals 85
Skylights 75
Slingers 220
Smoking 79, 80
'so far as is reasonably practicable' 35
Solvents 80, 111
Sound:
    power level 163
    pressure level 163
Space:
    confined 70
    working 71
Special wastes 155, 245
    consignment note 156
Sprain 210
Stain tubes 130, 150
Standards, British Harmonized (EN) 15, 17, 167, 176
Statutes 13
Statutory duty, Breach 6
Statutory instrument 14, 18
Strain 210
Stress 113, 123
Sub-contractors 191
Suitable person 117
Supervisor 47, 57, 98
Supporting structures 200
Supplier 91
Switches, isolating 231

Target organs 110
Temporary workers 25, 27, 28
Temperature 70, 194, 248
Tenosynovitis 113
Tetanus 113
Thackrah, Charles Turner 19
Tinnitus 161
Toilets 76
Tools:
    pneumatic 197
    power 197
Tort 6
Trade unions 29
Training 28, 35, 53, 57, 173, 210, 212, 222, 235
Transportable:
    gas containers 183, 186
    pressure receptacles 139
Travelators 76, 216
TREMCARD 143, 153
Trespass 6

Tribunal:
    Employment Appeal 10
    Industrial 7, 13
Trip devices 178
Trucks:
    driverless (robot tugs) 225
    fork lift 223
    pedestrian operated 224
    powered 214, 222
TWA (time weighted average) 149
Two-hand control 179

Underground:
    machinery 171
    services 197

VCM (vinyl chloride monomer) 131
Vehicles 194
Ventilation 70, 81
Ventilators 75
VDU (Visual display unit) 81
Vibrating tools 113
Vibration white finger 113
Volenti non fit injuria 67
Voting:
    positive procedure 14
    negative procedure 14

Walkways 74, 222
Washing facilities 77
Waste paper baskets 81
Waste management licence 247
WATCH (Working Group on the Assessment of Toxic Chemicals) 149
Water, drinking 78
Welfare facilities 35, 194, 198
Windows 75
Wiring, temporary 197
Women:
    child bearing age 26, 28
    pregnant 26, 28
Work equipment 173
Workplace 69
Workstation 72, 122
Work equipment 171
Writ 67
Writer's cramp 113
WRULD (work related upper limb disorder) 105, 113, 213

X-rays 114

Young persons 26, 28, 55